INVITATIONS
TO INVESTIGATE

INVITATIONS TO INVESTIGATE

An Introduction
to Scientific Exploration

PAUL F. BRANDWEIN

AND

HY RUCHLIS

ILLUSTRATED WITH LINE DRAWINGS
AND PHOTOGRAPHS

Harcourt Brace Jovanovich, Inc., New York

This book is based on the card set entitled *100 Invitations to Investigate* by Paul F. Brandwein and Hy Ruchlis, published by Harcourt Brace Jovanovich, Inc., in 1966.

FIRST EDITION
ISBN 0-15-238835-4
Library of Congress Catalog Card Number: 77-102440
PRINTED IN THE UNITED STATES OF AMERICA

Contents

List of Investigations

INVITATIONS
TO INVESTIGATE

CHAPTER 1

First Notes on the Art of Investigation

There was a time when the west was an open frontier. A boy or girl, dreaming of the future, could think of open lands to settle. There were new worlds to conquer.

Daniel Boone could go west and find the prairie and the forest, the field and the stream—free for the people who followed. Now there are people almost everywhere on this planet. The explorers of old have few places to go. The frontiers on this planet are almost closed. But the frontier of the mind was never more open. Man can search for new knowledge and explore new scientific discoveries. He pioneers in flights to the moon and to distant planets. He explores the region of the microscopic world, the atomic world, and, at the opposite scale, the giant galaxies and vast reaches of space. These frontiers are wide open for those who are trained in the skills of such explorations.

The new frontier is that of scientific investigation of the unknown. Each science has its own frontiers, whether it be of the oceans, the deep recesses of the earth, outer space, or the world of the small. But all explorers of this new frontier must have some things in common—the ability to investigate, and the understanding of the ways of those who penetrate the unknown and produce new knowledge.

But how does one learn to investigate? A simple answer might be this: one learns to investigate by investigating. A musician becomes proficient at playing an instrument by constant practice. A baseball pitcher becomes expert by pitching. A lawyer becomes skilled at his profession by handling many law cases. One learns by doing. No doubt we oversimplify when we say, "If you want to learn how to investigate, you should investigate—and in so doing learn how." And that's what this book is about.

What does the word "investigation" mean to you? Do you think of a detective dusting fingerprints or sending some particles of clay from a shoe to a lab for testing? Or perhaps you think of a committee of Congress meeting to find out the causes of poverty or the methods of improving education—in preparation for writing new legislation. Certainly, these are investigations.

Everybody investigates, and quite frequently. A businessman often investigates the possibilities of selling a product before offering it for sale. A high school senior generally investigates the desirability of attending various colleges before making out his applications. A housewife may investigate how different laundry bleaches clean clothes or how adding different spices affects the taste of food she cooks. Even a dog "investigates" a strange odor, in his crude but effective way, to find out who or what caused the odor. Scientists, of course, have raised the art of investigation to its highest levels, and it is this skill that you need to develop in preparation for your future.

The Nature of Investigation

In simplified form, an investigation is a search for answers to questions that interest the searcher. The question need not be stated but may be implied in his analysis or in the procedure he uses.

Simple questions of this kind occur very frequently in

your life. "What shall I wear today?" You investigate this question by looking out the window to see what the weather is like or by listening to a weather report on the radio or by asking someone who has already been outside.

Investigations are often easy to perform, involving only one or two observations and procedures before a reasonable judgment is obtained. But with the increasing complexity of our society, questions become more complex. The investigation of the pros and cons of candidates for President, for example, is generally very complicated. Some people may investigate such a question intensively, devoting hundreds of hours to reading, listening to candidates and commentators on TV, or discussing the matter with others. Some investigate very superficially or not at all, basing judgment solely on what party label the candidate may have. But whether the investigation is trivial or profound, the question remains a subject requiring investigation.

Questions to Investigate

What questions might you investigate? Any question in which you are interested can serve as the starting point. However, we do suggest certain areas of investigation in this book because they illustrate different procedures that are useful to investigators in various fields of science. If you are sufficiently interested to pursue any of these questions, well and good. But if you can generate your own questions and apply the principles described in this book to those investigations, so much the better. The important thing is to learn how to *perform* investigations and to *experience* the kinds of procedures scientists use, as well as to read about them.

Any fact, any observation, any speculation, anything you wonder about, can serve as the starting point. You needn't look very far. For example, continue to look at this page, but now shift your attention for a moment to the paper on which this book is printed. Why is paper selected for the page rather than plastic or metal? How was it made? What does it look like under a magnifying glass? What would happen if you were to heat similar paper in an oven at 300° F.? At 400° F.? Why does it burn if you put a lighted match under it? Why "under" and not "over"?

Or you can focus your attention on the letters, words, sentences, and paragraphs you see on this page, rather than on what they mean. There are many questions you could ask about these printed markings and their arrangement. What letters are used most frequently? How does the writing in this book differ from that in other books by different authors? Are the sentences longer, for instance? Why do the arrangements of letters have meaning for us? On page 28 of this book you will find An Investigation of Writing and Style, which indicates how one goes about investigating questions of that kind.

Suppose that you glance up from the book and observe someone sitting at a table. No doubt, you have done this thousands of times before. But now focus on questions about this observation. Why do people sit? Why does the chair have four legs? Why is the top of the table level? These are not trivial questions at all, and each of them can be pursued to great depth until one is soon unable to answer further questions without lengthy investigations.

For example, let's pursue the question, "Why do people sit?" An obvious answer might seem to be, "Because they become more tired when standing than when sitting." Well then, "Why does one become more tired when standing than when sitting?" Or one might wonder what would happen if a person sat all day. Is sitting always more relaxing than standing? One would have to find out how the muscles of the body react to the force of gravity, how one stands still, how the muscle cells respond when one stands or when one sits, how the nerves and brain and body chemistry all play a part when we "feel tired." As you can see, these are complex questions that could quickly lead into advanced scientific research.

Questions abound all around us. Glance at the wall. Why is it vertical? Look at the ceiling. Why is it about eight feet high in most houses, perhaps ten or twelve feet in a classroom, often thirty feet or more in an auditorium?

That clock on the wall—how does it tell time? How does your ability to estimate the passage of time compare with that of a clock? Does a clock tell time accurately on top of a mountain? In a deep mine? In a spaceship?

That window in your room—why can you see through the glass? Would ten sheets of glass noticeably reduce the light coming through? What would happen with a hundred sheets?

That brass knob on the door—is it magnetic? If it is magnetic, why does it have that property? And if not, why not?

That sound you hear outside—how did it reach you? Did it come through the wall, through the glass of the window, through the air, or by some other means?

As you can see, there is no limit to the number of subjects for investigation that one can generate. One could, in fact, list hundreds or thousands of questions with a modest amount of effort. If everybody were to do the same, and duplications were weeded out, billions and perhaps trillions of questions would be generated. Clearly, there must be some limitations on what questions to ask because there must be some time left to investigate. All questions, no knowledge, and no research make Jack a poor investigator.

Since time is limited and most of the questions are quite complex, we generally need to make use of man's most powerful tool of communication—the printed word—to help us investigate.

Using Reference Material

Books and journals are available that describe significant investigations others have carried out in the past. As time goes on and the work of each generation is added to that of previous ones, knowledge begins to accumulate rapidly. Each generation builds on the work of previous generations. It is foolish to overlook our vast heritage of knowledge, to repeat what others have discovered—unless, of course, we do so as a hobby or perhaps as a means of learning how to investigate.

Remember, then, that one essential element in most real investigations is the reading of books and journals related to the investigation. That is why such a high premium is placed on the ability to read.

But remember, too, that not all answers to questions are found in books and that not all statements in print are correct. Misinformation does abound—more so in some printed matter than in others. Some newspapers, for example, are notoriously inadequate for learning about social or political events. Others are reasonably reliable. Even when it comes to the less emotional scientific "facts," man's interpretations of the facts have a slippery way of changing as our knowledge grows. For example, it is interesting to look at a book about atoms published in the 1850's. You would find atoms described as unalterable in shape or form. Modern books describe atoms as whirling complexes of electrons, protons, neutrons, mesons, baryons, pions, and leptons. The position of a particle within an atom is not viewed as fixed in space at all. Instead, scientists refer to the likelihood (or probability) that a particle is within a given range of positions.

The Nature of Error

Even observations that seem perfectly straightforward and obvious are often subject to change for different observers, and also for the same observers under different conditions. For example, examine the photograph of the moon on the next page. Do you see craters in the picture? Let us see how accurate your interpretations are. Turn the book upside down. Strange! Mounds become craters and craters become

Photograph from the Mount Wilson and Palomar Observatories

mounds. The differences in the terrain of the moon that you seem to see are an illusion. They arise solely from the position of the photograph and our misinterpretation of the shadows.

Scientists and other learned men are not at all immune to error. The history of our concept of a flat earth versus that of a round one, or of a fixed earth versus one that revolves around the sun, shows how tricky "facts" are. Or we can cite a more recent example. Up to around the year 1940, doctors advised persons whose limbs had been frostbitten not to warm them too rapidly in warm water. It was thought best to rub them with snow and then warm the frozen limbs gradually. Actually, during World War II various methods were investigated with soldiers suffering from frostbite, and it was found best to warm the limbs rapidly with warm water. Not so long ago doctors advised against cooling burned parts of the body with water. Now, application of cold water and ice is advised in many cases.

Therefore, information in books must be viewed with some caution. Much depends on the source. A scientific journal is a much more reliable source of information than a popular magazine, although, no doubt, there may be times when a popular magazine would be right and a scientific journal wrong. Despite the chance of error in books and journals, they do contain vast numbers of reasonably correct answers to questions, and they do provide an enormous head start in answering many questions. In the past, undue reverence and almost worship of the words in books has caused widespread error to be perpetuated. Don't worship the facts in books, but do have a healthy respect for the knowledge books contain. Keep a questioning attitude and some caution with regard to the possibility of error in these books.

The nature of facts in print differs with the source as well as with the purpose. A lawyer, for example, reading the text of a law in a journal published for that purpose could be reasonably sure that it is a correct statement of the law— although rare misprints that distort meaning have been known to occur. After all, the *wording* of legislation is absolute, although its meaning may not be so. But a printed article by a scientist in a science journal has a different quality than a reprint of legislation. The work of the scientist may not be accurate, may be wrong in some respects, or may even be totally wrong. Thus a scientist must learn to read with a somewhat different approach than does a lawyer.

The Experiment

Investigations in science usually contain a special element —the experiment—that gives science its special character. In an experiment, a scientist designs a set of procedures to provide observations that will help solve the problem at hand. Planning is an essential element in the process. The planning of procedures may be simple or it may be complex, but it is always present. For example, suppose the question is,

"What is the most frequently used printed letter on this page?" The design of the experiment is simple—just count all the letters on the page and record how many of each kind there are. But suppose one asks, "What is the most frequently used letter?" Then the planning is much more complex. If a certain letter is found to be most frequently used on this page, is that also true of the next page? The next ten pages? Every page? If one letter is found to be most common for this book, is it most common for all books? Can one investigate *all* books in that respect? If not, how many books should one investigate? How many pages should one select in each book? How should the selection be made? If a particular letter is found most frequently used in English, is it also most frequently used in French? Spanish? Swedish? As you can see, experiments can be simple for simple questions dealing with a narrow range of events, but the questions can generally be broadened and made more complex by extending the range of the question.

Could one find out what letter is most frequently used just by thinking about the matter? Hardly! One might make a shrewd guess based on experience, or, in scientific terms, one may make a "hypothesis." But no one can know whether the hypothesis is probably true until the actual counting of letters has been performed.

Many books for young people list simple "experiments," generally where the information is already known. For example, you may already be aware that heating a material causes it to expand. If you read a description in a book of a procedure for observing the expansion of a steel bar and *actually perform the procedure, is it an experiment?* Not in the true scientific sense. Rather, we might call it a "demonstration" of a fact already obtained as a result of an experiment.

But suppose you decide to pursue the matter. You design a method for measuring small changes in volume of water with changes in temperature and proceed to heat some tap water. You find that it expands when heated. Then you try heating ice water at the freezing temperature. To your surprise the water now contracts when heated slightly. Spurred by this "discovery" and particularly by the contradiction to an established fact, you now plan and carry out a series of procedures in which you measure the change in volume by heating water from freezing temperature to boiling. You discover that water from 32° F. to 39° F. contracts when heated and thereafter expands.

Excited by your "discovery" of a contradiction to the "fact" that heating a material causes it to expand, you rush to your science teacher and disclose the startling news. No doubt he will refer you to books that describe, in greater detail and accuracy than you have achieved, a thorough picture of the expansion of water and its *exceptional* contraction with heating near the freezing temperature. Moreover, if you are interested and pursue the matter further, you will find much work done by scientists as to why such exceptions occur. Consequently, although you have discovered, or rather *uncovered,* a fact by yourself, it was really not new and therefore not a real discovery to the world. Although your experiment may be an experiment to you, it is not a significant experiment for the scientific world at this time in history. Perhaps we might call it a "little experiment"—one that has significance for you, if not for the world. You taste the thrill and excitement of discovery—or "uncovery"—and obtain an experience that is important in your development as a capable investigator.

In this chapter you have been introduced to the nature of investigations and to some of the pitfalls that you may encounter in performing an investigation. The next chapter pursues the matter further and provides some insight into methods of making investigations.

CHAPTER 2

Learning to Investigate

A First Simple Investigation

How would you find out a very simple thing—say, the meaning of the word *optimum?*

If you don't already know the meaning of the word, you would look it up in a good dictionary. To look something up is, in a sense, to investigate it, for to look something up is to find something out in the simplest sense of the phrase.

You wouldn't make up the meaning of a word. You would use the meaning agreed upon. Agreed upon by whom? By scholars who have studied the language. In a sense, then, when you go to the dictionary, you go to the work of scholars. Or in other words, when you go to a dictionary, you *consult authorities.* The reason: because authorities have spent their lives assuring themselves of the accuracy of their findings. It would be a mistake not to use this simple way of investigation, of finding out: *look it up.*

A Second Simple Investigation

Suppose you were interested in finding out what a *llama* is

—or if it is spelled *lama*. Perhaps you heard someone use the term or heard it mentioned on TV. You could ask others what a llama is. Some might tell you it is an animal found in Peru. Some might tell you a llama is a high priest of Tibet. Others might say they don't know. You could, of course, go to Peru or Tibet and find out for yourself. But you would agree this would be inconvenient. To find something out, one uses not only the most accurate source, but also a convenient one—at least at the beginning. The investigator tries to economize on time—but accuracy comes first.

Once more we go to a convenient source: books written by authorities. First to an unabridged dictionary. Then, perhaps, we go to encyclopedias and to articles written by authorities on the subject. By the way, is it llama or lama? Is a llama an animal or a high priest of Tibet? Or both? Or neither?

A Third Simple Investigation

You move to a new neighborhood. It becomes important to find out where most of the stores are—the grocer, the pharmacy, and the like. You could ask a policeman; or your mother or father or brother might help search them out with you.

But suppose you want to discover what birds are to be found in your neighborhood. Suppose, too, no one in your neighborhood has this information. Furthermore, you don't know one bird from another. Then, perhaps, you would need to try an investigation of your own. You might get a book with pictures of common birds. You would want to go out on many mornings and see which birds you can identify. You would make a list of the birds. Next year you would repeat your observations.

This is a simple investigation, very straightforward. But you will agree it isn't like finding out the meaning of a word such as optimum, or whether a llama exists, or what a llama is.

It is a kind of investigation a scientist might do—to be specific, an ornithologist (a scientist who studies birds) might do. Of course, an ornithologist would already know a great many birds, so he could list the birds in your neighborhood more quickly than you.

But this investigation, however straightforward, takes time, energy—and, furthermore, planning.

Toward a More Complicated Investigation

The procedures in the investigations we have described are simple. How shall you learn the ways of the investigator when faced with more complex problems?

You will learn best by doing. Why not proceed to do several investigations? We have selected three that require no elaborate equipment. Simple materials such as mirrors, books, papers, clocks, and the like, plus yourself, are all that you will need.

As you do these investigations, take two sets of notes. First, take notes of your observations. This procedure is important in *all* investigations because our memories are fickle. Observations recorded a day later—or even an hour later— are quite likely to be less accurate than if recorded immediately. Second, take notes of your method of investigation. Such notes will be important in deepening your insights into the ways knowledge can be obtained.

After you have done the three investigations, proceed to Chapter 3, which contains a more detailed analysis of investigating.

An Investigation into the Sensing of Time

BACKGROUND

We usually think of ourselves as having five basic senses —sight, hearing, touch, taste, and smell. However, we also have related senses—or perhaps we should say "combinations of senses"—that are very important in our sensing of the environment and determining our reactions to it. For example, we have a sense of balance that tells us whether we are standing upright or not. Other senses—related to the sense of touch—tell us whether it is hot or cold, respond to pressure exerted against the skin, and react to pain. We also have a kinesthetic sense, which keeps us aware of the positions and motions of our arms and legs and other parts of the body.

Our awareness of some aspects of our environment, such as the passage of time, involves a combination of several basic senses, as well as things we have learned. We can easily tell the difference between a one-minute interval and a one-hour interval. But just how accurate is our "sense" of time? Can we tell the difference between half a minute and one minute? Between fifty seconds and one minute?

Now, before reading further, try to plan some "little experiments" to provide information about this general problem. What procedures would you follow? How would you ensure accuracy? What measuring instrument or instruments would you use? Can you formulate several questions that extend and enlarge the scope of the investigation?

After you have done that, proceed with your reading for some clues as to how you might begin to investigate the way we sense time.

TRY THESE INVESTIGATIONS

1. How accurately do we sense passage of one minute of time? Ask a friend to tell you when one minute has elapsed. Be sure that he does not look at a clock or watch. Start him off when the second hand of a clock or watch is at the "12" position. Note the actual length of the time interval that he senses to be the end of a minute. Do not tell him whether he was accurate or not—or how accurate.

Repeat this several times. How accurate is his judgment? Is he always wrong in a similar way, or are the estimates sometimes too small and at other times too large?

Try this with different people. Do they differ in ability to estimate one minute of time?

2. Can people learn to sense time accurately? Repeat the previous investigation. But this time, after each trial, tell the subject what the actual length of time was. Does he learn after several trials? Does he remember what he learned after a week or two?

3. What intervals of time are judged most accurately? Try the investigation for periods of ten seconds, one minute, ten minutes, and one hour.

4. Do different conditions affect the sense of passage of time? Try the experiment while a person listens to music, or watches a movie, or performs a task of some kind. Try tapping him on the shoulder. Make loud noises. Does he judge time differently if something is happening? Does the type of event affect his sense of time?

5. Invent your own questions and design experiments to gather more information.

An Investigation into Mirror Writing

BACKGROUND

Place a sheet of paper in front of a mirror. Stand a book up in front in such a position that you cannot see the paper directly but can see its reflection in the mirror. Now try writing your name while watching the handwriting in the mirror. Be sure not to look at your hand. Did you find it very difficult to write your name? How do you explain the results?

When you write, you usually watch your handwriting as it appears on the paper. Through experience you have learned to make the eye and hand muscles work together. But when you watch your writing in a mirror, everything looks reversed. Objects and places that are really nearer than others appear farther away. As a result, what you see appearing on the paper (your handwriting) seems to have no connection with what you feel your hand doing. The nerve impulses from your eye and your hand that are relayed to your brain do not seem to be interpreted as they normally are. The habits formed through the years seem no longer to work properly.

Does this initial investigation arouse your interest? Do

you want to know more about it? No doubt, you can think of additional problems and "experiments" by yourself. Before you read further, try to formulate some questions and design some experiments to answer the questions.

TRY THESE INVESTIGATIONS

1. Draw a set of lines on a sheet of paper as shown below. Try tracing each line from the center while viewing it in a mirror.

Are certain lines easier to trace than others? Which ones?

And, by the way, what is the point of using a set of standard lines rather than handwriting?

2. Do people differ in ability to trace lines in a mirror? Are certain lines easy to draw for some people, difficult for others? Design your own experiment.

3. How long does it take for different people to complete tracing a triangle drawn on paper while looking in a mirror? Are some people quicker at this task than others?

4. How long does it take to learn to trace a given figure such as a triangle viewed in a mirror? Can you measure progress in some way from the tracings? Keep records of the time required and see how you improve with practice. Do this over a period of a week or more. Do you ever reach a point at which your mind is completely adjusted in the new situation and you can trace a figure as easily in the mirror as without it?

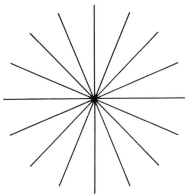

5. Try to write in reverse, so that it looks correct when viewed in the mirror. How long does it take to learn to do this? Do people differ in this ability?

6. Type a paragraph using carbon paper, inserting the carbon backward so that the copy comes out on the back of the sheet you are typing. The writing is reversed, the way it appears in a mirror. Test different people to see how rapidly they can read this reversed writing. Is it related to ability to do mirror writing?

Can this reversed carbon copy be used as a quick test to discover which people are good at mirror writing? You may also want to measure rates of learning how to read or write this way.

7. Invent your own questions for investigation and experiments to answer the questions.

An Investigation of Writing and Style

GENERAL BACKGROUND

Some persons claim that Shakespeare was not really the author of the great poems and plays for which he is world-famous. They believe that another man living at the time, Christopher Marlowe, was the real author and that Shakespeare was merely a "front."

How can you test such a claim? One way would be to examine the published works of both men and compare their writing styles. Such a study has been made, and to some persons it indicates that the published works of Marlowe and Shakespeare were written by two different people. How could the investigators come to such a conclusion?

During his lifetime, each person develops certain habits —he walks in a certain way, talks in a certain way, eats in a certain way, and writes in a certain way. These and other characteristics are so individual that we can recognize a person by his appearance, by his voice, or even by the sound of his footsteps. Banks recognize the signatures of persons on their checks. You can easily learn to identify the handwriting of your friends if you are interested enough to observe samples of their writing.

In a similar manner, each person writes letters and compositions with an individual style. He uses certain favorite expressions, begins and ends his sentences in a characteristic way, has a tendency to make sentences and paragraphs long or short. One person may write in the present tense most of the time while another prefers the past. Certain common words like "so," "therefore," "because," "and," etc., are used in different ways by different people. It has even been found that the average number of letters in a given number of words varies from writer to writer.

A SAMPLE INVESTIGATION

Here are passages of approximately 100 words, from two different authors. The first is from *The Planets* by Franklin R. Branley, and the second from *Math Projects: Map-Making* by Tannenbaum and Stillman, both published by Book-Lab, Inc.

Author A

A distinctive characteristic of Jupiter is the pattern of dark bands that appears all around the planet, which are generally parallel to the equator. The explanation of the dark bands has not been made. They may be produced by vast dust clouds suspended in Jupiter's atmosphere, or they may be produced by ice crystals. Perhaps neither of these possibilities is correct, and the dark bands may be produced by something entirely different. The details of the surface are ever-changing; therefore, there must be considerable turbulence in the atmosphere of the planet. The shift of the surface markings is fairly uniform| at times, which suggests there may be prevailing winds.

Author B

A map is a special kind of picture. There are maps that tell about heights of mountains and depths of valleys. Others show boundaries of states and nations. Road maps show distances and directions between places. Globes often give a picture of the whole Earth. Sky maps show where to look for stars and planets. All maps use symbols to tell their stories. Symbols are signs. Each symbol has its own meaning. For example, the red and green traffic light is a symbol you see every day.

On a map, each symbol tells you something, too. When you understand the| meaning of the symbols, you can read the map.

(*Note:* The vertical line in the last sentence of each passage indicates the end of the 100th word.)

How can we analyze the writing style of each author? A critic, of course, would analyze the styles in his own special way and describe how he feels about each. But then, critics will differ and sometimes find themselves at opposite extremes. However, if all we want to do is show that styles are different, we can introduce the scientist's favorite tool—measurement—into the picture. For example, let us look at the number of paragraphs each author uses, the number of sentences there are in each passage, and the average number of words per sentence. Are there any words or phrases that each author uses several times? You may observe the following differences in the paragraphs.

Author A uses the word "a" once, while B uses it five times.

Author A uses only one paragraph. Author B uses three.

Author A uses six sentences in the passage with an average number of words per sentence of 18.2, while author B uses twelve sentences with an average number of words of 8.8.

Author A uses the phrase "may be" four times. On the other hand, author B uses the word "you" four times.

Author A begins three of the six sentences with "The." Author B doesn't begin any of his twelve sentences that way.

Now let us make a study of the lengths of words in a total of 100 words by counting up all the words with a given number of letters, counting a hyphen or an apostrophe as a letter. This is what we find:

Number of Letters per Word	Number of Words	
	Author A	Author B
1	1	5
2	21	10
3	22	27
4	11	17
5	7	13
6	5	8
7	11	14
8	8	0
9	5	3
10	3	3
11	2	0
12	1	0
13	2	0
14	1	0

You can readily observe that author A uses a greater number of long words, on the average. In the passage of 100 words he uses 22 words of 8 letters or more, while author B uses only 6. With regard to words of medium length (4 to 7 letters), author A uses 34 while author B uses 52.

Of course, these passages are too short to make definite conclusions about the writings of the authors, but the kind of analysis made here could be applied in a similar way to complete books or to several books by the same author.

TRY THESE INVESTIGATIONS

1. Make an analysis of the style of writing of several authors. Find out how they differ. Then ask a friend to copy a short paragraph by each of the authors, without showing you which book it was taken from. Identify the author.

2. Study compositions by several of your friends. Then ask each to write a paragraph or two, using capital letters to disguise the handwriting. (If they can type their paragraphs, so much the better.) Can you identify each writer?

3. Extend your study to identification of handwriting. Work out a system for describing the way each person writes. Then use it to identify written samples.

4. Obtain sample signatures from several friends. Then ask each to do his best in copying the signatures of all the others and to turn in a test sheet with all the copied signatures, plus his own signature. Using the original signatures as your guide, pick out the true signatures from the copied ones.

5. Are there any basic elements common to all writing styles? Letters, of course, are the basic stuff of which words are made. All authors use 26 letters, no more, no less. But do different authors favor certain letters?

It has been found that in all writing of any length, the letter *e* is the most commonly used letter, averaging about 13% of the writing. The letters t, a, i, o, and n are also common. Letters like z, q, and j are, as you know, much less frequently used.

Make a study of frequency of use of letters in several different books, if possible, 1,000 letters or more per selection. Is the frequency of occurrence of letters the same for all writing? What differences occur, if any?

You may wish to apply this knowledge to solving cryptograms. This is an interesting puzzle hobby that depends upon knowledge of frequency of occurrence of letters and also on your knowledge of patterns of letters in words.

6. Invent your own questions to investigate, and design experiments to gather information.

CHAPTER 3

Toward an Art
of Investigation

What did you learn from the three investigations described
in Chapter 2?

First, it is clear that an investigation can arise from anything you observe or from any interesting event. The problems chosen for investigation were of a kind that are not
limited to subjects like biology, chemistry, physics, or mathematics. They begin with some real event or observation or
question—one that is, above all, interesting. In other words,
the whole world and all events in it are subjects that may be
investigated.

Second, each problem tends to grow in many directions
as you work on it. For example, in the analysis of writing, it
was suggested that you investigate lengths of words. Quite
readily such an investigation leads to another on lengths of
sentences and paragraphs. The question as to how often
each letter occurs is then quite natural. As one gathers information about any of these questions by counting and recording, the mind soon begins to probe in other directions.
The letter *e* is the most frequently used letter in all writing.
How often do the other letters occur? Is the order of occurrence the same, or very similar, for all writing? Is the information found for English the same as for French or Bur-

mese? Is the pattern of words and sentences the same for spoken language as for written words?

In this way a chain of questions or problems can arise, with one suggesting another, then another, and so on.

Did you notice how experiments were designed in each of the investigations? In each case, the overall pattern of the experiment is suggested by the question or problem. For example, suppose the question is: Do people differ in ability to trace figures while looking into a mirror? Obviously, one answers the question by observing different people attempting to trace figures while looking in a mirror. Of course, the important details of the experiment remain to be worked out—but the general pattern is clear. If one asks: Can people learn to judge time?, then an experiment concerned with timing a number of persons for a number of trials immediately suggests itself.

It is for this reason that it is often said that the proper statement of the problem is half the solution. The question, or statement of a problem, often suggests what kinds of experiments must be done to obtain answers.

The procedures of counting, measuring, and recording the information are fundamental in solving scientific problems. For example, how does one find out whether author A uses more complex words than author B? We proceed to count letters in words and record the data in a manner illustrated in the table on page 31. Once the *doing* part of the experiment is completed, anybody can study the table and easily compare the numbers. He could arrive at the same conclusion that you do.

How can we compare the ability of two people to trace a figure in mirror writing? A measurement of some kind is required. We could measure the time required to trace a figure. Or we might measure the accuracy with which it is traced—perhaps by measuring how far off the line the tracing occurred. We might do both.

How would we compare the ability of people to judge time? The measurement in this case is quite simple—we just note the time at which a person thinks a minute (or some other interval) has elapsed. If in his first trial he judges a minute to be ten seconds and by the tenth trial is consistently judging it to be fifty-five to sixty seconds, then it is clear that he has learned to judge approximately a minute's time. Without the measurement of the interval of time, it would be impossible to tell whether a judgment is close or not. It is easy to compare the ability of a number of people this way.

Why is it that these investigations were not difficult? Certainly, they don't compare in difficulty with an investigation into the chemical composition of the atmosphere of Venus, or of the Rh factor in human blood. But the investigations that are easy for you wouldn't be simple for a five-year-old.

An investigation begins with a certain state of knowledge of the investigator. At the start he has certain concepts that are part of his previous learning. (At the end of the investigation, the investigator may have learned something and developed additional concepts. Perhaps these concepts will be transmitted to others by means of discussion or by publication of papers.)

What is a concept? Below are drawings of two groups of

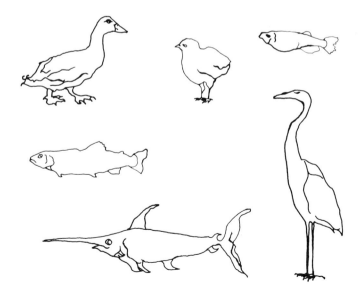

animals. Classify them into two main categories, a task that should pose no difficulties.

No doubt you have grouped them into the categories of birds and of fish. Why? You have a clear concept of a "bird" as an animal with feathers and wings, two legs, a specific shape, and other characteristics.

You have a clear concept of a "fish" as an animal that has scales, fins, a specific type of shape, and other features. Development of such concepts requires considerable observation, learning, or investigation. Unfortunately, many errors are possible in development of concepts, and individuals— often mankind—may obtain concepts that do not conform with reality. An investigation may then be necessary to correct the concept or replace it.

As one works on an investigation and gathers information from experiments, certain clues are provided that suggest ways of obtaining new facts. For example, one could ask, "What is the second most frequently occurring letter in the English language?" This suggests that data be gathered by studying a variety of writings. Suppose that at the completion of analysis on the first piece of writing, it turns out that the letter *t* is the second most frequently occurring letter. This suggests that it may be true in general. But, of course, since only one piece of writing has been examined, we certainly are not entitled to stop there and declare it a fact that *t* is actually the second most frequently used letter.

This is a very simple example—and actual investigation in science would be more complex. But you might picture the general procedure of a scientific investigation as follows: a scientist, having spent his life in learning, begins his work with a considerable degree of understanding and knowledge. Knowledge and understanding are always incomplete or even, at times, incorrect. Nevertheless, a scientist begins a piece of research with an understanding—a concept (or group of concepts). Assume that a scientist (many years ago) was working in the field of prevention of disease; specifically working on the disease pellagra. A basic concept he brought to his work was: Many diseases are caused by bacteria. On that basis it was reasonable to state the hypothesis: A bacterium causes pellagra. However, after considerable effort no bacterium could be found as a cause for pellagra.

In this type of situation, a scientist will seek other hypotheses. It was known that some diseases are caused by lack of a vitamin. So the hypothesis was changed to: Pellagra is caused by lack of a vitamin. This hypothesis suggested a different direction of research and design of new experiments. Eventually, this direction proved fruitful—lack of a vitamin was discovered as a cause of pellagra and this, in turn, led to methods for its cure.

The diagram on page 38 shows the general pattern of scientific investigation. If you study the diagram, you will once again come to realize that the scientist does not really begin with a problem. He begins with long study and training; he learns what others before him came to know and how they came to know it. In other words, he learns of the orderly explanations developed by other scientists; and he learns, through investigation, how they tested their explanations. On this basis he can begin his own investigation because he can clarify a problem. On the basis of the *known*, he investigates the *unknown*.

Furthermore, what he knows can also lead to a working hypothesis—an educated guess as to a possible solution to the problem. Sometimes a working hypothesis is not too difficult for a scientist to make, although the approach to a solution is difficult. For example, when it was first observed that the planet Uranus did not seem to move in the sky exactly as predicted by Newton's Laws, several astronomers made the hypothesis that this deviation was caused by the force of gravity of an unknown, more distant planet pulling Uranus off course. But making the calculations and predicting where the new planet would be were the real obstacles in this case.

A diagram of a Scientist's Way: his methods of intelligence

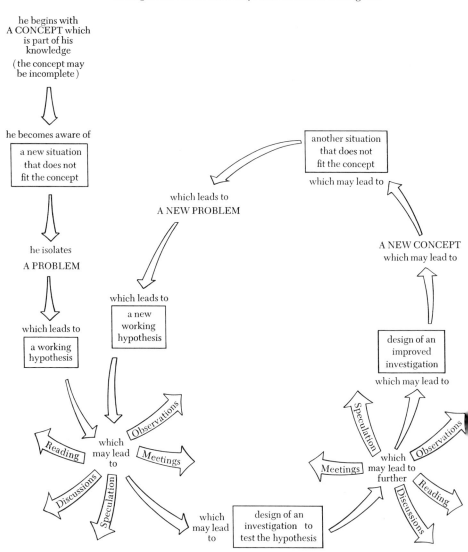

Some of the greatest discoveries, of course, were made by scientists whose hypotheses were so different that they were revolutionary in concept. Such was Pasteur's hypothesis that some diseases are caused by germs; or Einstein's hypothesis that the observed speed of light was not affected by the motion of the observer.

Once a scientist has a hypothesis (or set of hypotheses) for his problem, he investigates by learning as much about it as possible through reading, discussion, speculation, direct observation, and meetings with other scientists. At some stage along the way—often after prolonged investigation—the design of an investigation to test the hypothesis emerges. A key element in most scientific investigations is the design of *experiments* to gather data regarding different aspects of the problem.

An experiment is that part of an investigation in which the investigator seeks to "control" the environment so as to produce observations and data which can answer questions that are posed. An investigation that illustrates the nature of an experiment is included at the end of this chapter.

In complex investigations there often is a series of experiments to provide observations and data. As the experiments are performed, information begins to emerge that supports or rejects the hypothesis—or sometimes provides no clues whatsoever. Occasionally, an accidental observation suggests an entirely new problem, which may cause the scientist to drop his old problem and work on the new one. Frequently, flaws develop in the experiment, and the scientist may wish to refine or alter his experiments. He may sense the need to know more about minor details, and so undertake new side investigations that enlarge his knowledge and provide a firmer base for tackling his main problem. The result of years of such activity may finally lead to a solution to the problem and the development of a new concept, which, when reported in journals, adds to man's fund of knowledge. On the other hand, the investigation may lead to naught; that is a risk every scientist takes. But often, even when it seems to lead to nowhere, there are bonuses in the form of new knowledge about important details or techniques and increased experience for the scientist and his team.

Does the process end with a solution to the original problem? By no means. The development of a new concept, by adding to man's store of knowledge, provides a new base from which the vast unknown can be attacked. Every new concept suggests new problems to solve and opens new avenues of investigation. So the cycle of investigation continues —or perhaps we should call it a spiral because the scientist completes the cycle at a higher level of knowledge than when he began. Again there are new discussions, meetings, readings, speculations, and observations. New information and data give rise to a new problem, and the next round in the cycle begins.

The scientist's ways of obtaining new knowledge, which we have just described, are sometimes called the methods of the scientist or the processes of the scientist. Sometimes they are called the methods of inquiry. Percy Bridgman, a great scientist who won the Nobel Prize, called them the "Methods of Intelligence."

Now let's apply the "Methods of Intelligence" to a simple investigation and observe how even a simple situation can lead to error if the environment is not controlled during an experiment.

An Investigation into Bread Mold

BACKGROUND

Suppose that you want to investigate the nature of bread mold, and to do so, you need to grow this organism. After all, bread mold is a living plant.

It's simple to grow bread mold. Take a piece of white bread and leave it in the air for half an hour. The spores of bread mold are in the air, and some of them will fall on the bread.

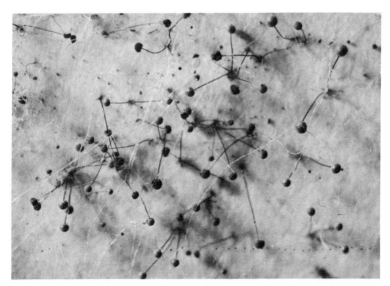

Put a few drops of water on the bread and put it in a jar that is kept in a warm place. In a few days you should find a growth of bread mold that appears as black or gray (or other colored) blotches on the bread.

You may begin to wonder whether there are ways to raise a better crop of bread mold. It would be necessary to observe how different factors affect the growth of the mold. You know that green plants cannot grow without sunshine. Perhaps bread mold, too, grows better in the light than in the dark.

To find out, you might take two pieces of bread and divide each into two equal parts. Add ten drops of water to each piece of bread, then place the pieces in different jars. Put two jars in a dark closet and two in sunlight.

This was done by one student, and he found that more bread mold developed in the jar kept in the light. On the basis of this investigation, could we conclude that to grow a better crop of bread mold it should be exposed to light? Let's look into this a bit more closely.

Was the temperature in the closet the same as the temperature in the sunlight? Suppose it wasn't. Suppose the temperature was higher in the light. This could be true, couldn't it? After all, light energy can be changed to heat energy.

It might be, then, that the greater growth in the light was not because of the light, but because of the higher temperature in the sunlight.

To do this investigation properly, then, we must be sure to control the temperature so it is the same in the light and in the dark.

What was wrong with our first investigation? Scientists would say that the investigation was not well planned because not all the variables in the experimental conditions were controlled. As you can see, failure to control all the variables is one of the great pitfalls in doing an investigation.

CONTROLLING THE VARIABLES

Light and temperature are not the only conditions that affect the growth of a plant. The plant must be watered, too. Suppose you want to find out whether moisture is also necessary for the growth of bread mold.

Start by toasting two pieces of bread to a crisp and exposing them to air. Boil water to kill any mold spores that might be present in it. (Why is this important?) Put ten drops of the boiled water in a jar that has been sterilized with boiling water. Keep the other clean jar dry. Put a piece of dry bread in each jar, and keep them both on a table side by side. Cover both jars. Then the conditions of temperature and light would be the same for both jars.

In this investigation, moisture is the variable. It is the condition of moisture that varies. One jar varies from the other only by having moisture in it.

If mold grows only in jars without moisture and does not grow in jars with moisture, you might say that bread mold did not need moisture for growth or that moisture inhibited growth.

If bread mold grows only in jars with moisture and not in jars without moisture, you might say that bread mold needs moisture for growth.

You would be able to say this because there was only one difference between the jars. The other variables, such as light and temperature, were the same—they were controlled. In our investigation the variable being investigated is the presence of moisture.

In doing any investigation, then, you must search out all the variables. You must be sure that only one variable is being studied at a time.

Of course, the investigation should be repeated at least several times to be sure that the result was not accidental or that some time factor was not operating to influence the results. Before scientists accept the results of an investigation, they prefer to have a number of different persons perform the procedure. If everyone obtains the same results, then they can be much more confident of its correctness.

What can we learn from the procedures in this investigation? First, note that it was essential to "control the variables" in each situation so that we could be absolutely sure that our observations differed only because of the difference in the one variable. We are reminded of the sick person who takes some pills, stops working, gets plenty of sleep, and eliminates heavy meals. He gets better. Well, then, did the pills cure him or was it the fact that he stopped working or got more sleep or stopped eating heavy meals; or was it even the normal recuperative processes that operated over a few days or weeks? One cannot be sure. It would require a carefully controlled experiment with a number of people suffering from the same illness to be sure of the effectiveness of the pills.

In the remainder of this book, there are numerous suggestions for investigation of interesting problems. Perhaps you can apply the "methods of intelligence" to these problems and thereby climb a bit higher along your path to new knowledge.

In each of the chapters to come, you will find emphasis on one or more aspects of investigation. For example, Chapter 8 emphasizes *measurement*. Nevertheless, you will find that other aspects of investigation—such as *observing* (detailed in Chapter 5), *gathering data* (detailed in Chapter 7)—are also utilized.

Any investigation calls on all "methods of intelligence" to one degree or another. These methods are useful in many things you do—not only in the investigations in this book.

CHAPTER 4

An Investigation—
Its "Limits"

What excites your curiosity? Why do you investigate some things and not others? Why are you interested in one *phenomenon* and not another?

What is a phenomenon? One dictionary defines it as "something visible or directly observable, as an appearance, action, change, etc., . . . any unusual occurrence, an inexplicable fact." In other words, a phenomenon is any event or object or observation that arouses your curiosity, and is therefore a subject for investigation.

As you may have noticed in Chapter 1, phenomena exist all about you. They can be simple or complex. You may look at a horse standing on four legs and then look at a man standing on two and ask interesting questions about the phenomenon of standing. You can listen to some music and ask questions about why you like it or don't like it . . . or why some people like it, while others don't. On the other hand, some phenomena cannot be observed and studied without highly specialized equipment, as is the case with pulsating stars and galaxies, or interactions of nuclear particles.

In this chapter we present several investigations that involve some simple phenomena. Perhaps you will be able to determine—in some small way—what *limits* there are to your investigation of these phenomena. Or are there *no* limits?

An Investigation into the Sense of Touch

BACKGROUND

You may have read of Helen Keller, who was blind and deaf and seemed doomed to be forever shut off from other people. Through the devotion of a dedicated teacher, Anne Sullivan, Helen was taught to speak and even to "hear" speech by using her sense of touch. This achievement indicates how remarkable this sense really is.

Close your eyes and touch several different objects—a book, a pair of scissors, a glass, a piece of paper, a wooden ruler. It's usually easy to tell what you are touching. You can feel the texture of the material and the shape of the object. In that way you can identify it.

Different parts of your body differ in the ability to distinguish materials and shapes. For example, try identifying materials and objects by touching them with your wrist or elbow or toes. Are these parts of the body as effective as fingers in identifying the object?

A simple investigation suggests why these differences in sensitivity may exist. Tape two pencils together so that the points of both rest on the table when the pencils are held vertically. Close your eyes and touch the two points to-

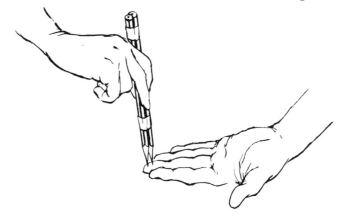

gether to a fingertip. Do you feel both points? Now try the same thing on the elbow, wrist, and toes. Do you feel two separate points—or do they seem like one?

In the fingers, there are many more nerve cells sensitive to touch than there are in other parts of the body. For that reason, we can make better judgments about the nature of objects with our fingers than we can with other parts of the body.

TRY THESE INVESTIGATIONS

1. How do different parts of the body differ in ability to distinguish one point touching the skin from two points touching the skin?

A wire hairpin is a useful instrument in this investigation because it can be easily adjusted to different distances between points by bending. Begin by setting the points one-quarter of an inch apart.

Ask a friend to cooperate in your investigation. Gently touch various spots on his arms, legs, and back with the two points of the hairpin at the same time. At what places on his skin can he tell that two points are touching? At what places does he feel only one?

It is important that the subject not see the hairpin when you touch his skin. Use a blindfold to make sure that information is obtained only from the sense of touch. Also, at random intervals during your investigation, touch the skin with a single point. In this way you can be quite sure that a person's answer is not influenced by what he expects to feel. If he is always correct in his judgment, you may assume that he can really tell the difference between one and two points. If he is wrong as often as right, you may assume that he can't tell the difference.

Now try increasing the distance between the points of the hairpin. Again test different parts of the body.

From this information can you tell which parts of the body are more sensitive than others?

2. Test different people. How do they differ in this respect?

3. Make a selection of different objects, and test different people as to their ability to distinguish size and shape by touching the objects with different parts of the body.

4. Paste squares of different materials, all the same size, on cards. Can blindfolded people tell the difference between the materials by the sense of touch?

5. Invent your own investigations of the sense of touch.

An Investigation into the Heartbeats of Small Animals

BACKGROUND

One of the first things a doctor does when he examines a patient is to take his pulse. The pulse is a "beat" of the artery at the wrist; it indicates the rate at which the heart beats.

It would be interesting if the pulse could be seen with the naked eye. But if this is generally not feasible with human beings, it is possible to see the heartbeats of some small animals such as the earthworm (*Lumbricus terrestris*) and the water flea (*Daphnia pulex*).

In the water flea, the heart itself can be seen beating. You will need to examine the animal under the low-power lens

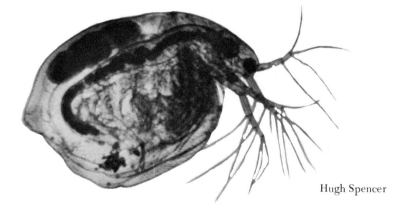

Hugh Spencer

of a microscope. The normal heartbeat of *Daphnia* is very rapid, but with practice (and a little inventiveness) you will learn to get a pulse count in beats per second.

The earthworm's dorsal blood vessel is visible to the eye. So is its pulse. To understand the unusual characteristics of the earthworm's circulatory system, you will need to know something about its anatomy and physiology. It would be most helpful to spend some time in library research with reference books and encyclopedias.

Earthworms, of course, are readily available in the ground or in fishing supply stores. *Daphnia* is generally available in stores that sell aquarium supplies—these organisms are used as food for fish.

TRY THESE INVESTIGATIONS

1. What is the pulse rate of the earthworm or of *Daphnia?* Do different individuals have the same pulse rate?
2. Does the pulse rate depend on the size of the animal?
3. Does the pulse rate change during the day or night?
4. How does the pulse rate respond to different temperatures?
5. What is the effect of different colors of light on the pulse rate? (Clue: Cover a flashlight with cellophane or plastic of different colors.)
6. What is the effect of an increase of carbon dioxide? (You may have to design a chamber in which the concentration of CO_2 can be varied.)
7. Think of other interesting problems to investigate.

An Investigation into the Resolving Power of the Eyes

BACKGROUND

The eye may be thought of as an optical instrument, and all optical instruments have certain limits to their ability to make different details visible. For example, look at the black-and-white photograph of a group of people below. It has black areas, white areas, and many intermediate shades of gray. Was gray ink, as well as black, used to print this photograph? If so, were there four or five shades of gray ink?

Examine the photograph with a magnifying glass. Now you can see that the gray areas break down into separate black and white dots. The dark gray areas have larger black dots and smaller light ones. The light gray areas have smaller black dots and larger white ones. It is clear that the grays in the photograph are not produced with different kinds of gray ink but by printing small dot patterns that are partly black and partly white.

An enlarged pattern of the boxed area in the photograph is shown on page 51. View the enlarged pattern from a distance and observe how the black dots blend into a fairly clear picture.

The unaided eye lacks the *resolving power* to distinguish the tiny black-and-white dots at a distance. Instead, it sees the dots blended together, giving the appearance of continuous areas.

Harbrace

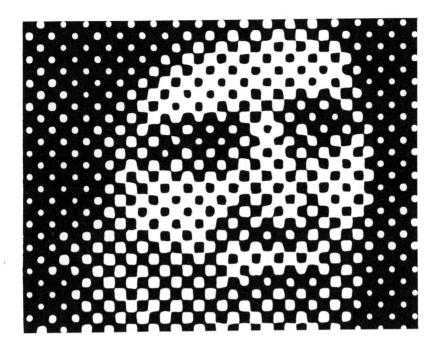

In the same way dots of only four colors—such as black, red, blue, and yellow—used in the proper proportions, can be made to blend into many desired colors. Thus a full-color picture in a magazine, with all its different shades, tints, and hues, is produced by printing the page four times, each with a different colored ink.

TRY THESE INVESTIGATIONS

1. Examine different *printed* photographs with a magnifying glass. Place a ruler on the dot patterns and count the number of dots per inch. How many dots per inch do you find for newspaper photographs? Do newspapers and magazines differ in the number of dots? Are there any methods of printing that do not show dot patterns when photographed?

2. How many dots per inch do you find for photographs in books printed on dull paper? How many for photographs printed on very glossy paper? Is there any relationship be-

tween the kind of paper used and the spacing between dots?

3. Look at a regular photograph (not a printed one) with a magnifying glass. Does it break down into dots the way newspaper and book photographs do? If not, examine it with a microscope under different powers. Do you see separate dots or particles? Do negatives differ in this respect from photographic prints? Is a "fine grain" film different from a regular one?

4. Test different persons as to ability to see the dot patterns in newspaper photographs without magnifying glasses. Can some persons see the dot patterns without the aid of a lens? How do older persons compare with younger ones in this ability?

5. Measure the ability of different persons to resolve dot patterns. Obtain sheets of graph paper with different spacings. On each sheet, ink in the alternate spaces to make a checkerboard pattern of at least 20 by 20 squares, as shown below. Mount all the patterns on a large board and illuminate them brightly.

Have each subject stand quite far away, at a distance great enough for the largest dot pattern to appear continuous. As he slowly walks toward the board, let him announce the place at which each dot pattern begins to appear. Mark

off and record the distances. Set up a measuring scale for the ability of the eye to resolve separate dots. What is the variation among persons? Is it related to age? Is there a relationship between the size of the pattern and the distance at which it becomes visible?

6. How does the ability of persons to resolve small objects differ with illumination?

7. The dot pattern of newspaper, magazine, or book photographs may be used to measure very small distances. Place a thin object, such as a nail or wire, on a printed photograph, and using a magnifying glass, count the number of dots that it covers. Then count the number of dots per inch. From this information you may calculate the thickness of the object.

8. Formulate your own problems and design investigations to gather information about the problems.

As you look back at the three investigations of phenomena described in this chapter, do you see the points of similarity in approach in each case? Observe how the original situation leads one to ask a variety of questions that are subject to further investigation.

Were you puzzled by an apparent lack of hypotheses? They were there in every case, but disguised a bit. For example, in the investigation of the pulse rate of small animals, various questions were raised. Some questions can be restated as hypotheses. For instance, one question was: "How does the pulse rate respond to different temperatures?" We can formulate this question as the hypothesis: Temperature affects the pulse rate. Another question was: "What is the effect of different colors of light on the pulse rate?" This question can be formulated as the hypothesis: Light of different colors causes changes in the pulse rate.

Note that before data is gathered, one cannot tell in advance whether the above statements are true or false, nor can one know to what degree any of these factors affect pulse rate. But as you make observations and begin to test the hypothesis, new directions open up. For example, suppose you find that increased carbon dioxide content in the air around the earthworm causes an increased pulse rate. This raises the question as to whether this is true for dogs or rats or even man. Further questions arise about the exact relationship between the amount of carbon dioxide and the pulse rate. What is the pulse rate if the air contains 1% carbon dioxide? 2%? 5%? 10%? What other gases increase the pulse rate?

As you can see, the boundaries of an investigation of any phenomenon are practically limitless and restricted only by the knowledge and *ability* of the investigator and the time he has available—and his *resources*. What resources are necessary to carry on an investigation? How do they limit an investigation? These and other practical considerations determine the extent to which investigations are actually pursued. But in many cases, interest on the part of the investigator is the major factor which determines the outcome. An interested investigator will often find the time and resources to pursue an investigation to a successful conclusion.

CHAPTER 5

Observing

Everybody knows what "observing" is . . . but does each person know *how* to observe? Observing is an art that can be mastered only by practice and training. A good detective looks for and observes clues that others never see. A doctor observes a person's face and behavior in a way that is very different from that of the man in the street. A scientist's way of observing is very much more complex than that of a layman.

In this chapter are included several investigations that require careful observation of details that one would ordinarily miss. Try these for some practice in the art of observing.

An Investigation into How Droplets Form

BACKGROUND

Have you ever noticed droplets of water forming on a glass of ice water in the summer or on a window in very cold weather? The droplets start so small that they appear as a mist, but gradually they grow larger until they run down the side of the glass. A study of the way these droplets form and grow makes a most interesting investigation.

It is best to have the droplets form on a horizontal surface so that they don't slide down and spoil others that are forming. A good way to do this is to cover a container of water, such as a glass or plastic dish, with a piece of clear glass or plastic. (It is important that the water be kept in an enclosed space so that it does not evaporate entirely over a period of time.) Set the glass in a warm place—perhaps in a sunny window or above a warm radiator. As the water evaporates, some of it condenses on the inside of the cover, forming beautiful round droplets like those in the figure.

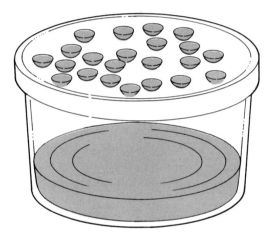

Before the droplets are easily visible without a magnifying glass, you may observe what at first appears to be a mist. With a magnifying glass you can see that it is really composed of very tiny droplets. These slowly grow in diameter.

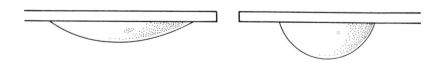

Although the droplets are pulled downward by gravity, they do not fall, because they are also pulled upward by a force or attraction exerted by the solid material of the cover. This attractive force is known as adhesion. It is different for different solids and liquids, so that the way the droplets form depends upon the materials used. For example, a plastic cover will form droplets at a different rate and of different size and shape than a glass cover.

The condition of the surface also affects the way droplets form. Water (and other liquid) molecules attract each other with a force known as cohesion. A water droplet is able to stick to a solid surface because the adhesive force exerted by the molecules of the solid is stronger than the cohesive force of the other water molecules. The solid then pulls the liquid to itself, so that the droplet tends to wet the surface. If the opposite were true and cohesion were stronger than adhesion, the water would not wet the surface or even stick to it. For example, when a surface is oily or dirty, adhesion is weakened. Then the water no longer wets the surface, and no drops form on the underside of the cover of the jar of water.

TRY THESE INVESTIGATIONS

1. How do water droplets form on different materials?
2. How does the temperature of the water affect the way the droplets form?
3. In what pattern, if any, do the droplets form on various materials?
4. Do bigger droplets "gobble up" smaller ones as they grow? How does this vary for different materials?
5. Is there any connection between the size of a drop and the amount of clear space around it?
6. What is the effect of using soap or detergent in the water? What happens when salt, sugar, or other materials are dissolved in the water?

7. What is the effect of vibration on the formation of droplets? (You can obtain a vibrating motion by placing the glass on or near some electrical device, such as a refrigerator or a fan.)

8. How do droplets form when the surface is covered with oil, talcum powder, or other impurities?

An Investigation into Rainbows

BACKGROUND

You have seen rainbows arching up into the sky. They are beautiful to look at, and they make an interesting study for investigation.

Rainbows may be visible in the sky if both the following conditions are met:

    a. The sun is low in the sky (that is, the time of the day is late afternoon or early morning).

    b. The sun's rays are reflected from droplets of water (A, B, and C in the drawing)—usually from a rain cloud.

If you stand with your back to the sun, you may see the reflected light from the raindrops forming a rainbow. If the

rainbow is very bright, you may be able to see a larger but fainter circle of colored light surrounding it. This is called a secondary rainbow.

TRY THESE INVESTIGATIONS

1. Make a study of rainbows. Every time you see a rainbow, observe it carefully and keep a record of your observations. Note the time of day, weather conditions, the shape of the rainbow, and the way the colors are arranged. Do you see a faint secondary rainbow outside the main one?

What happens to the rainbow if you walk a distance to the right or left of it? Where is the sun in relation to the rainbow?

Where does the rainbow end?

2. On a sunny day, you can make your own rainbow by using the fine spray from a garden hose. Face away from the sun and locate the shadow of your head. Direct the hose so that a strong spray of water surrounds the shadow. You may see a circular rainbow with its center located at the shadow of your head. How does this rainbow compare with the kind you observe on a rainy day?

3. A halo, or ring, sometimes appears around the sun or moon. Observe the colors in such a halo. How does the halo differ from a rainbow?

4. Make color photographs of rainbows and study them carefully.

An Investigation into Variations in Living Things

BACKGROUND

You have heard the expression, "as alike as two peas in a pod." The peas are, of course, very much alike, but if you examine the individual peas carefully, you will see differences in size, color, shape, arrangement of parts, etc. Such differences become important in many types of scientific investigations. For example, Darwin's view of evolution depends in great measure on the seemingly small but very significant differences between individuals of a species.

Sunflower seeds (sometimes sold as "polly seeds") in grocery stores are quite suitable for a study of variations because of the wide variety of white and black stripes on the seeds and the many different shapes. Other kinds of seeds can also serve for this purpose.

Insects are particularly good subjects for investigations of this kind because it is easy to find large numbers of a single species. Some suitable choices would be Japanese beetles, potato beetles, cabbage butterflies, or sulfur butterflies (sometimes miscalled yellow cabbage butterflies). All of these are spotted, striped, or marked in some way. If you collect enough insects, you can find specimens with a full

Charles E. Mohr, National Audubon Society

range of markings. The variation may be so great that at one end of the range the specimens may seem to belong to a new species or at least a new variety.

Obtain a book that describes methods of pinning insects and spreading them out so that you may examine their variation. The way of collecting them depends on the kind of insect you choose to study: beetles can be picked up by hand, while butterflies require a net.

Whatever insect you work with, you will need a way to kill it quickly. You may want to make a special killing jar out of a widemouthed jar in which you place rubber bands soaked in carbon tetrachloride. A piece of blotting paper will keep the rubber bands confined. (Since carbon tetrachloride is poisonous to insects, you may expect it to be poisonous to other living things—including you. *Be sure that you do not breathe in any of the vapors.* When you open the jar, do so out of doors or near an open window.)

TRY THESE INVESTIGATIONS

1. What variations do you find in a given type of seed? Among the characteristics you may wish to study are: length, width, weight, shape from different points of view, general color, patterns (stripes, bands).

2. Select a particular characteristic such as length or weight. Measure this characteristic for at least a hundred seeds. What is the range of variation? Find a way to show the variation by graphic means.

3. Investigate the types of variations in different characteristics of a species of insect. How much variation do you find?

4. Investigate variation in beans, flowers, grass, trees, and other forms of plant life.

5. Investigate variations within people of similar age, for example; eye color, length of arm or fingers, skin color, shape of ear, etc.

An *Investigation into Cloud Development*

BACKGROUND

Watch a cloud on a sunny day as it drifts lazily across the sky. How peaceful and quiet it looks! But actually the air in many clouds is in rapid motion. Small clouds often grow quickly into big ones and generate rain, hail, lightning, and thunder. We do not notice the motion because the clouds are so far away.

A cloud forms when air is cooled so much that the water vapor in it condenses into droplets. The type of cloud depends on how the air is cooled and the way the air is moving.

Some clouds are formed by air that rises, expands, and cools. Air close to the ground may rise because it is warmed when the ground is heated by the sun. Air may also rise as it passes over a mountain or is pushed by cooler, drier air moving into the region.

Most clouds formed by rising air are lumpy and billowy in shape, due to their upward movement. These *cumulus* clouds usually have flat bottoms, which show where the rising air begins to get cold enough to condense into droplets. Small cumulus clouds, which form on a warm summer day, often grow steadily until they form big, rainy storm clouds called "thunderheads," or *cumulonimbus* clouds. The clouds shown in the picture on the next page are of the *cumulus* type.

Another way in which clouds are formed is by air moving over a cooler surface. Clouds formed in this way are the flat, sheetlike *stratus* clouds, which cause large-scale overcast, fog, drizzle, or steady rain.

If the cumulus or stratus clouds are fairly high, we refer to them as *altocumulus* or *altostratus*. Other basic types of clouds are *cirrus* (very high, thin clouds) and *nimbus* (heavy rain clouds). These names can be combined in several ways to describe other types of clouds, such as *cirrocumulus, cirrostratus,* and *nimbostratus.*

U.S. Department of Commerce, ESSA, Weather Bureau

TRY THESE INVESTIGATIONS

1. Observe the development of fluffy cumulus "fair-weather clouds" on a clear day when the sun heats the ground. Watch a particular cloud for about an hour, studying its growth and changes in form. Does it get higher or wider? Does it rise or fall? Does it show any signs of evaporating or disappearing?

2. If you live in a hilly area, you may be able to watch clouds develop and disappear within a short period of time. As air travels over a mountain ridge, clouds tend to form on the side where the air rises and cools. As the air goes down the other side, it warms up, and the clouds tend to disappear. Take notes on the "life histories" of several individual clouds.

3. Find out the significance of the various types of clouds for weather prediction. Make weather predictions based upon cloud changes and check them afterwards.

4. Do you ever travel by plane? If so, on a long flight you can see a sequence of cloud types in a few hours that would normally take a day or more to pass overhead. Can you identify the cloud types when you observe them from above?

5. One good way to study the changes in clouds (from the ground or from an airplane) is to take photographs at intervals of about one minute. Sometimes you may be able to trace the cloud development of a cumulus "fair-weather" cloud up to the point where it grows into a cumulonimbus thunderstorm cloud.

6. Some home movie cameras are equipped to take pictures at timed intervals. If you have such a camera, take one photograph per second on the film, and later project it in the normal way at sixteen frames per second. In this way you can observe things happening at sixteen times the normal speed. Seen in this way, clouds appear to roll and boil, with violent air currents suddenly revealed. Take photographs of different types of clouds and study the changes in them. (You may have to experiment a bit to find a suitable time interval between pictures.)

An Investigation into Shapes of Living Things

BACKGROUND

The skeleton of a sea urchin is shown on the next page. We observe first that the overall shape is circular. Second,

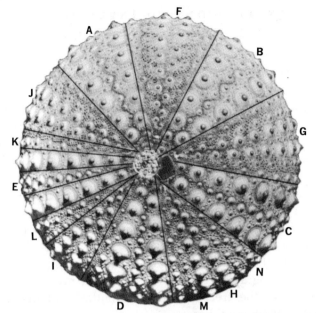

we see that the skeleton contains many small mounds. Each mound has a cone-shaped base with a small hemispherical bump on top. These bumps line up into a pattern of ten main radiating arms, as in the photograph. The arms alternate, with five arms (A, B, C, D, E) containing larger bumps, and five arms (F, G, H, I, J) containing smaller ones. Within each arm we can also detect still smaller radiating arms (K, L, M, N, etc.). The five-fold symmetry displayed in this double pattern of five identical arms reminds us of the five arms of a starfish. As a matter of fact, the sea urchin is closely related biologically to the five-armed starfish.

Examine the photograph of the sea urchin in greater detail. No doubt you will note other regularities, particularly in the grouping of the bumps.

Similar studies of external pattern and shape can be made for many other kinds of living things, both plant and animal. Observations made in such investigations play a basic role in classifying living things and determining their origin.

TRY THESE INVESTIGATIONS

1. Make a study of the shapes and patterns of seashells.

2. Study the shapes of leaves. There are several different varieties of maple leaves. Although they have certain similarities, they also differ in some ways. What are the basic similarities in the leaves of all maples? In the leaves of the horse chestnut and its relatives? In the leaves of ash trees? In other kinds of trees?

Study the outline shapes of the leaves and also their branching vein patterns.

3. Study the overall shapes of trees. For this purpose it is best to photograph trees out in the open, rather than in a forest where the trees crowd one another and distort the shapes of their neighbors.

You might select one basic kind of tree, such as the maple, and compare shapes of different varieties of maple. Or you might decide to study the basic shapes of completely different varieties (elm, birch, oak, hickory).

4. Investigate the shapes and patterns of flowers. Different parts of a flower have their own regularities.

5. Observe the patterns in shapes of insects and other animals. The patterns in the wings of butterflies and moths and the surface markings of beetles are quite interesting.

As a result of your experience with observations in these investigations, what lessons did you learn with regard to method of observing. Can most observations be made quickly? Do they require planning? Can you remember all observations or should you keep useful notes? Are measurements helpful? What sense (or senses) are used most often in making observations? What role can photographs play? Are any observations unexpected? Can they lead to hypotheses and new questions for investigation?

In some ways careful observation is the most important of

the many skills required in scientific investigation because it is the basic raw material from which everything else follows. An investigation based on poor observation is generally worthless. But an investigation based on careful observation can usually stand on its own. Other scientists can then make use of the observations to extend and develop the investigation further.

CHAPTER 6

Observational Errors

Ask a very young child how big the full moon is. You may get the answer that it is as big as a plate or a pie. You know, of course, that the moon is about two thousand miles in diameter. It only appears to be small because of its vast distance from us—about a quarter of a million miles.

Similar considerations are true of everything we observe. For example, in the first investigation in this chapter, you will see that most people grossly misjudge weights of fluffy, large objects. A wide variety of optical and other illusions affect our judgments about what we think we observe. Scientists, too, are not immune to observational error. Around 1900 one scientist thought he had discovered a new type of radiation which he called n-rays. These n-rays could only be observed in total darkness. Many papers were written by other scientists reporting different characteristics of these difficult-to-observe n-rays. Eventually, it was found that all of the observers were wrong—there were no n-rays at all. They were merely "seeing" figments of their imaginations in the dark.

In this chapter are described several investigations that probe different types of observational error. Perhaps, after doing these, you will devise your own investigations in this fascinating subject.

An Investigation into Sensing Weight

BACKGROUND

Imagine two objects of about the same weight—one large and filled with air spaces, the other very compact and dense. Repeated trials show that almost everybody would judge the first object to be much lighter than the second when placed in the palm of a hand. Some people would even judge the compact one to be more than ten times as heavy. Why are the judgments inaccurate?

An object placed in the palm is pulled downward by the force of gravity and so tends to squeeze the skin a bit. Since the weight of the more compact object is concentrated on a very small area of the skin, the "pressure" under the object is high. The nerves under the skin respond by sending a signal of "increased pressure" to the brain. The brain interprets this to mean "heavier."

The weight of the less compact object is spread out over a wide area of the palm. The pressure on any one spot on the skin is much less, and furthermore, the skin is dented very little. As a result, the signal from the nerves to the brain is weak and is interpreted as "less pressure," or "lighter."

TRY THESE INVESTIGATIONS

1. Make a collection of ten objects such as a Styrofoam block, a sponge, a light bulb, an empty box, a lead sinker, a small metal wheel, an opaque vial filled with sand, a rock, etc. Weigh each and keep a record of the weights. Try to select objects in a fairly narrow range of weights, with the heaviest not much more than about twice as heavy as the lightest.

Ask your friends to judge the weights and to place the objects in order of weight from lightest to heaviest.

How accurate are their judgments? Are some people much more accurate than others?

2. Place the objects in transparent plastic bags and have each subject judge the weights by holding them by strings attached to the bags.

How do the judgments now compare? Are they different from before?

3. Repeat the previous experiment with the subjects blindfolded. How do the judgments of weights compare with the previous trials?

4. Can people learn to judge weights by feel, after some experience? Use a scale to measure the actual weights of a number of objects in the same weight range as your original ten test objects. Let each subject handle the objects to learn as much as he can about how weights of different sizes and materials feel. Then see how accurately he can judge the weights of the ten completely new test objects.

An Investigation into Optical Illusions

BACKGROUND

Which line shown below is longer, AB or BC? Or are they the same size? Guess. Then measure them with a ruler. Are you surprised at the result?

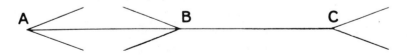

What you have observed is an *optical illusion*. Let's find out more about how your eyes can be fooled.

We obtain information about our surroundings by means of our senses of sight, hearing, touch, taste, and smell. Information, in the form of electrical signals, is sent to the brain from our sense organs—located in the eye, ear, skin, tongue, and nose. The brain then interprets these signals. But the interpretation depends on what has happened before—what you have learned and experienced. If your brain receives a signal similar to one it has received before, it may interpret the signal in the same way. For example, if you should hear a bell-like sound produced when someone strikes a goblet with a spoon, your brain might interpret the sound as that of chimes produced by a visitor who presses the button at the front door. This would be an *illusion:* a signal from the senses that is wrongly interpreted by the brain.

What causes the optical illusion in the drawing? It is easy to see that the slanted lines have something to do with it. Suppose you investigate more precisely the way in which the slanted lines cause the eye to be fooled.

TRY THESE INVESTIGATIONS

1. Draw different versions of the optical illusion on cards. In some drawings make AB equal to BC. In other drawings make AB greater than BC, and by various amounts.

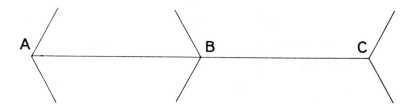

Vary the angle of the slanted lines as shown above. Make some almost parallel to AC and others almost perpendicular to AC.

Number the cards in a random way. Then show them to a friend and ask him to state for each one whether AB is equal to, longer than, or shorter than BC. Keep a record of his answers. Do the same with other people.

Does everyone agree on the judgments, or do people differ? How does the angle of the slanted line affect the judgments?

2. Investigate what happens to the illusion under the following conditions:

    a. if AB and BC are very long
    b. if AB and BA are very short
    c. if the slanted lines are short
    d. if the drawing is colored
    e. if the lines are dotted instead of solid
    f. if the lines are white and the background dark

On the basis of this information devise a hypothesis to explain why the illusion occurs.

TRY THESE MORE DIFFICULT INVESTIGATIONS

1. Lines DE and FG in the drawing on the next page are straight, as you can see by placing a ruler alongside. The radiating lines seem to have the effect of pushing outward the sections of the horizontal lines that are nearest to the center.

Investigate this illusion by making several variations of the drawings on cards and testing the reactions of your friends. Vary the spacing of the parallel lines and also of the radiating lines. Change the outer shape by making

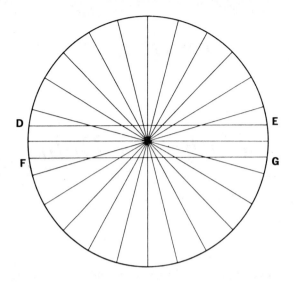

it a square, a very tall rectangle, or a very flat rectangle.

Vary the darkness of the lines. In one drawing make the radiating lines light (or dotted) and the parallel lines dark. Reverse this arrangement in another drawing. Try different colors for different types of lines.

Devise a theory to explain why this illusion occurs.

Can you construct a drawing in which the lines seem parallel but are actually curved?

2. The moon seems larger when it is lower in the sky. Is the image of the moon really larger near the horizon, or is this an illusion? Take photographs of the moon when it is low in the sky and again when it is higher. Check your conclusion by:

a. viewing the moon upside down on the horizon.

b. making several artificial moons of slightly different size and viewing them in a dark room at different levels.

From these investigations would you say that our senses are completely reliable—or completely unreliable? Can we trust our senses at all? What precautions might we take to prevent the types of observational errors indicated in this chapter?

CHAPTER 7

Gathering Data

In most investigations observations would be quite useless unless the observer kept careful notes and gathered *data* (singular: datum) that would subsequently be useful to him and often to others. Our memories are notoriously fickle. A day or two (or ten, or a hundred) later, the memories of most events become dim. We may even remember incorrectly, as you no doubt know from your own experience.

Scientists don't trust their memories when it comes to gathering data. They record it immediately. Thereafter, they may study the record and base calculations on it. But without such gathering and recording of data, most scientific investigations would be fruitless.

The investigations in this chapter feature skill in gathering data. As you perform each investigation, select appropriate ways to gather data and decide suitable types of data to gather.

An Investigation into the History of a Tree

BACKGROUND

At Mesa Verde in Colorado, there are numerous abandoned dwellings used by Indians who lived there centuries ago. Scientists who have studied the area tell us that about 1276 A.D. the Indians began to move away—or perhaps died out. There were no written records of this event. How do the scientists know when the Indians disappeared from the dwellings?

They know from a careful study of the growth of rings of trees. The Indians used trees as rafters and floor beams in their houses. A study of the rings of these trees shows changes in climate during their lifetime. When these rings are compared with those of old trees recently cut down, the changes can be matched and the ages of the ancient rafters calculated. It was observed that at a certain date all construction ceased. It is reasonable, then, to infer that this date marks the approximate time when the Indians disappeared from the region.

How can tree rings provide information about dates? In spring and summer a tree grows taller and wider as its branches are extended. The trunk and branches also grow thicker. The bark cracks, making room for the greater thickness as new interior cells are produced. As autumn approaches, the growth slows down and finally stops. The next year the process begins again. But there is a distinct mark inside the tree that separates the new growth from that of the year before. When a tree is cut down, the trunk shows a series of rings, one representing each year.

During some years growth is rapid because there is plenty of sunlight and rain, and no disease or insect infestation. At other times, however, growth may be slower. Perhaps the season was dry, or the tree was shaded by a larger one, or insects ate the leaves, or disease attacked the tree. The good years produce wide rings. The poor years produce narrow rings. One can tell by looking at the width of each of the rings which years were good and which were poor.

You can see these rings clearly in the photograph on the next page. At first, in region A, growth was slow, as shown by the narrow rings. Perhaps the tree was shaded by bigger trees and did not have much light. Or perhaps the roots had not yet penetrated deep enough to get a good supply of water. Then came a period of rapid growth (B). Suddenly

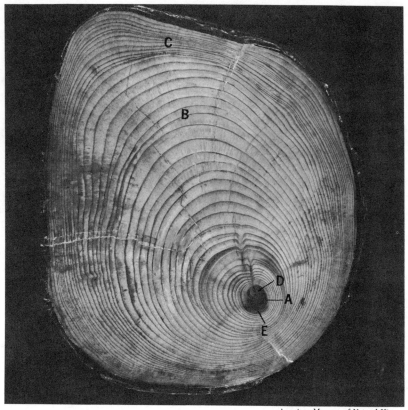

American Museum of Natural History

growth slowed down again and the rings became narrower (region C).

From the rings we can also learn about something else that happened during the life of this tree. The rings at the very center of the tree (D) are round. But a little farther out they become oval, with their earliest most rapid growth in the direction toward E.

This uneven growth indicates that the tree was probably bent over, perhaps by a strong wind—or perhaps by a nearby tree falling on it and causing it to lean over. You can see that a study of the rings can reveal a great deal of the history of a tree.

However, in such an investigation, keep in mind that a tree is an important living thing and that you shouldn't go around chopping down good trees. First, be sure to get permission from the owner of the land to cut selected trees. Then, select trees that are either dead or obviously dying. Mark on the tree the sections you wish to study. *Then obtain the help of an adult with sawing equipment.* Never chop or cut down trees by yourself. A falling tree can cause injury or death.

Perhaps an area is being lumbered, and you can get an adult to help you cut small sections from the lumber. You may also be able to get sections of trees at a sawmill.

If you locate recently cut tree stumps, you might photograph the rings for later study. In such cases, measure the diameter of the trunk so that you can later determine growth from the photograph.

TRY THESE INVESTIGATIONS

1. Measure the diameter of each ring of a section near the base of the tree. Then make a graph to show how the tree grew in width. From this graph you can quickly tell which were the periods of good growth and which were the periods of poor growth.

Observe any evidence of irregular growth. What is the nature of the irregularity? Can you think of reasons for it?

2. Compare the life histories of several trees of different diameters in the same region. Do they show similar periods of good or poor growth? Consult the local weather bureau or agricultural experts to find out whether they have records of dryness, insect infestation, forest fires, or other causes for poor growth in that area.

3. If there is a new forest near your home, study a recently fallen or dead tree and find out how old the forest is. Try to find out if the forest developed from an abandoned pasture or farmland. Was there a forest fire? Did lumbering occur in the area? Do your tree rings confirm these events? Can you determine their dates?

4. Study the history of an old forest by examining the

stumps of very large trees that have been cut down, perhaps for lumber. How old are the big trees?

5. Compare different kinds of trees as to rate of growth. Is there a pattern in the way a certain type of tree grows? Does it grow faster when young or when mature?

An Investigation into the Growth of a Plant

BACKGROUND

Plant a seed. The seed germinates, and soon the plant stem grows tall.

But which part of the stem grows fastest? The lowest part? The tip? The middle? Or do all parts of a stem grow at the same rate of speed?

You will agree that if you are going to investigate which part of a plant stem grows the fastest, you will need some growing plants. One very useful plant to use for this kind of investigation is the lima bean.

To germinate beans, you may use two clay saucers (the kind used to hold flowerpots). Cut a blotter to fit the bottom of one saucer. Then place the other saucer on top. Let the saucers sit in an aluminum baking dish half full of water, as in the drawing. The water from the dish will soak into the clay saucer and keep the blotter moist.

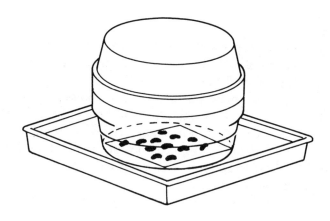

Now soak a dozen lima beans in water overnight. In the morning, place them on the blotter in the saucer. As soon as the roots poke through the covering of the seed, plant each bean separately in garden soil in a flowerpot. The best planting depth is about half an inch below the surface. Be sure to keep the different conditions (the variables) the same for each pot.

Water the pots each day, enough to keep the soil moist— not wet. It would be well to add the same amount of water to each pot. Keep the pots away from sources of heat such as the radiator. When the plants are about two inches tall, you are ready to begin your investigation.

TRY THESE INVESTIGATIONS

1. Make marks on the stems of the selected plants, spacing the marks every quarter of an inch from the surface of the soil toward the tip of the growing plant. Use India ink or any marking that will not wash off.

Between which two marks will the greatest growth take place? What is your best guess?

Examine the plants every day. How will you measure their growth? How will you record the data? Will a graph be helpful?

Do all the plants show similar results? If they do not, how do you account for the differences?

2. Gather data by making daily photographs of the plant stem. These would show where most of the growth took place. Do different types of plants differ in the rate of growth?

3. Design a procedure for investigating the growth of roots. Try out the procedure.

4. Are the results the same if the plants grow in the dark?

In the investigations in this chapter, it was necessary to make careful observations and record the data in some way. One simple method is to make a suitable measurement, in which numbers may be recorded. It is far superior to record a measurement such as "2.5 inches" than to say, "It's about as tall as the height of my little finger." A table of measurements then provides information that anyone can use, whether or not he did the experiment.

Photographs are another important means of recording data. In fact, in some sciences, such as astronomy, it is by far the best method. Photographs of the earth taken from satellites promise to revolutionize data gathering for weather prediction.

Graphs are extremely useful in analyzing and interpreting the data. A graph is a picture of the way one quantity changes in relationship to another. If one looks at raw data—usually in the form of numbers in a table—it is generally too difficult to see how a change in one quantity affects another. The picture provided by a graph is very useful in this respect.

In the next chapter you will explore the role of measurement in greater detail, by means of several investigations involving simple measuring instruments.

CHAPTER 8

Using Simple Measurements

Some measurements are performed with instruments that are very simple. For example, length is usually measured with a marked stick—a ruler. An angle is measured by means of a marked disk—a protractor. The amount of liquid in a container is obtained with a measuring cup or graduated cylinder.

A weighing scale is a bit more complex, consisting of springs or levers and frames to hold these parts. Nevertheless, the reading of the measurement on a dial is generally quite simple.

The mechanism of a clock is substantially more complex, consisting of a train of gears, a spring motor, a rotating balance wheel, and other parts. But again the actual measurement, from the position of hands on a dial, is simple enough.

In this chapter, it is suggested that you perform several investigations in which these relatively simple measuring instruments play a key role. But the procedures are quite different from customary usage—and will reveal some of the many ways in which measurements can be useful in making observations and gathering data.

An Investigation into the Measurement of Time

BACKGROUND

You can't put your finger on time, as you can with an object. You can't stretch out your arms to measure it, as you would a distance. Time seems to flow from past events into present events into future events. It is always changing, never still.

Yet we do have a sense of time, and it is important to measure it to conduct our everyday affairs. It is a most important quantity in many scientific investigations.

Watches and clocks are so common today that we take them for granted, but it was only relatively recently that methods for the accurate measurement of time were developed.

Suppose that there were no customary clocks in our environment. How would you then measure intervals of time?

Any regularly changing event can be used to measure time. The alternation of day and night is one such event. It is easy to count days. Years can also be counted by observing the changes in seasons. Nevertheless, it took many thousand of years of observation of the seasons, and of the motion of the sun and stars, before our present calendar of about 365¼ days in a year was developed.

Shorter intervals of time—hours, minutes, and seconds— must be measured in different ways. In some of his experiments, the famous scientist Galileo (1564–1642) used the regular dripping of water from a small hole in a container to measure time (see the diagram on the next page). Once he used the beats of his pulse (reasonably regular in an unstressed situation) to time the swinging of a chandelier in a cathedral. In this way he discovered a basic characteristic of the motion of the pendulum, which has since become the basis for pendulum clocks. Such a clock makes use of the fact that if a weight hanging from a rod or string of a certain length is set into motion, it takes the same time to complete every swing.

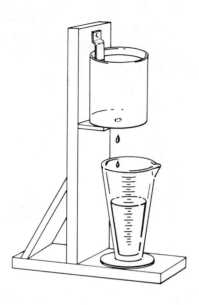

Most of our clocks and watches use the regular vibrations of springs to measure time. A simplified form of this method is shown in the illustration on page 85. When the weight is pulled down a bit and then let go, it vibrates up and down. Unless the spring is stretched beyond its limits, the time of its vibration does not depend upon the amount of pull. Most clocks and watches make use of spiral springs, with a small rotating wheel serving as the weight.

TRY THESE INVESTIGATIONS

1. Suspend a small weight—a heavy washer, for instance —from a string about three feet long. Pull it slightly to one side and let it swing. Count the number of swings in one minute. Now pull it farther over to one side. Again count the number of swings in one minute. Is it the same as before?

Determine the time of each swing in seconds. With this method, you can achieve an accuracy of about 1/100 of a second.

Now use this pendulum to measure the time of other

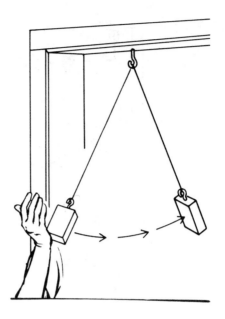

events. For example, you could measure the time it takes for a person to walk a block by counting the number of times your pendulum swings as he covers the distance. If you measure the length of the block, you can also calculate his speed.

2. Is the time required to complete one swing of a pendulum affected by the amount of weight on the string? By the length of string? If so, in what ways?

3. Using a spring, develop a similar method of measuring time. Suspend a weight from a spring. Pull it down a bit and let go. Count the number of vibrations in one minute. Then pull it down a little farther and let go. Count the number of vibrations. In each case, what is the time, in seconds, for a single vibration? Calculate the time to within 1/100 of a second.

Use this method to time some events. For example, let a small stream of water flow from the faucet into a glass. How much time is required to fill the glass to overflowing? Do this several times. Is the time always the same? Could you use a fine stream of water as a clock?

4. Is the time of vibration of a spring affected by the amount of weight on the spring? By the length of the spring? By the force used to pull the spring? If so, in what ways?

5. Use the information from the investigations in number four to detect slight differences in the weights of similar objects.

An Investigation into Rolling Liquids

BACKGROUND

Could you make a clock using honey?

It's not too difficult. Fill a jar half full of honey—or any other liquid that flows very slowly. Place the jar on its side on a board that is tilted slightly, as shown in the illustration on the next page. Hold it there for about a minute until the honey settles down to the lowest position. Then let go.

Watch the jar roll down the incline. You will see that it moves slowly at a steady rate. It may take several minutes or hours, depending on the positions of the jar and the

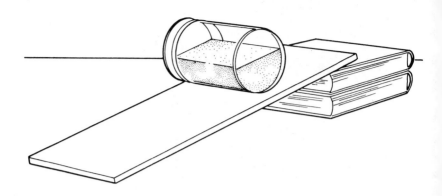

board. Since the jar takes the same time to roll one inch as it does to roll the next inch, you could use it for a clock.

In carrying out the investigations that are suggested below, you will need two basic measuring instruments for gathering data—a ruler and a watch or clock.

TRY THESE INVESTIGATIONS

1. Raise the high end of the board in ⅛-inch steps. For each position, measure the time it takes the jar to roll down the incline. Keep a record of the measurements. Then you can use this information to measure intervals of time.

A method of marking the jar and board to obtain more accurate measurements is shown in the illustration below.

2. It is interesting to try different amounts of honey in

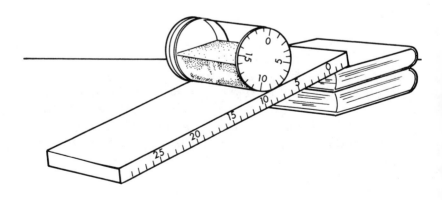

the jar. Does the jar roll faster or slower with more honey? Can you determine a relationship between the amount of honey in the jar and the time required to roll down a given incline?

3. Investigate how other liquids in the jar affect the way it rolls. Try molasses, rubber cement, oil, water. What kinds of liquids allow the jar to roll fastest? Slowest?

4. In general, heat makes liquids flow more readily. How do you think this would affect your honey clock? Test your honey clock outdoors on a cold day. Try it after you have cooled the jar in the refrigerator. Try it on a hot day or in a warm boiler room.

Was your prediction correct?

5. The time required for a jar to roll down a slanted board can serve as a measure of the viscosity (resistance to flow) of the liquid. If the rolling time is the same for two different liquids (other things being equal), then their viscosities are the same. If the jar rolls faster when filled with one liquid than when filled with the other, then the viscosity of the first liquid is less.

Compare viscosities of different liquids in this way.

6. Watch the action of the honey as the jar rolls. Can you explain why the jar rolls the way it does?

7. Improve the accuracy of your measurements. Since the measurement of time depends upon the position of the jar, some way of accurately locating the jar on the board is required. A simple way to do this is shown in the illustration.

An Investigation into Measuring Craters on the Moon

BACKGROUND

The photograph on the next page shows the moon when half of it is illuminated. Observe the many large, round craters and steep mountain walls that give the surface a pock-marked appearance. You can also see enormous dark areas that look like seas, but we are quite sure they have no significant depths of water.

Lick Observatory Photograph

Astronomers tell us that some of the craters are over 100 miles in diameter. How do they know?

It is possible to measure the diameter of a crater from a

photograph. First, use a ruler to measure the width of a crater on the photograph. Then compare the width with the entire diameter of the moon. For example, consider the crater at A in our photograph. Measure the widest part of the crater. It appears to be about ⅛ of an inch across, while the entire moon appears to be 6 inches across.

Astronomers tell us that the moon is 2,100 miles in diameter. (You will have to assume that fact for this investigation.) Since the diameter of the photograph is 6 inches, each inch represents 2100/6, or 350 miles. If you measure a crater to be ⅛ of an inch across at the widest part, then its actual diameter is ⅛ x 350, or 43.75 miles.

You will find it easier to measure distances with a ruler marked in centimeters than with one marked in inches. Centimeters are divided into ten equal parts, while inches are usually marked off in eighths and sixteenths. Distances in centimeters are therefore expressed in decimals that are easy for calculation, while fractions of inches require clumsy arithmetic.

It is a good idea to use a powerful magnifier to help measure small distances accurately. The proper way to use the magnifier is to bring it close to the photograph and then bring your eye close to the lens. Move head and lens up and down together until you get a clearly focused view.

TRY THESE INVESTIGATIONS

1. Select some of the largest craters in the photograph and measure their diameters. Then calculate their diameters in miles. Which is the largest crater visible in the photograph?

Note that there is a problem with the craters near the edge of the moon. They appear oval because of the slant of the ground away from us. How should your measurement be made to eliminate that factor?

2. Determine the sizes of other features on the moon: for instance—seas, ridges, and valleys.

3. Obtain a large map of the moon perhaps from a museum store or a map store. Measure the diameters of all the craters named on the map. Make a list of craters in order of size. What is the average diameter of all the craters that you measured?

An Investigation into Creeping Metals

BACKGROUND

Did you know that metals can "creep" and that they can also become "fatigued"? Of course, these expressions don't mean quite the same thing when applied to metals as they do when applied to people, but there are similarities.

All metals, and all materials for that matter, are subject to change over a period of time. As the metal parts of a machine (for example, of an airplane) are exposed to vibrations or even to the chemical action of air, they gradually weaken. Investigations of some airplane crashes have revealed that critical metal parts have become "fatigued" because of forces to which they have been subjected or because of vibrations of the airplane; then they gave way.

One way in which this condition occurs is by a gradual lengthening of a stressed metal part. The metal is said to "creep."

Creep is difficult to observe in strong metals because it requires a very large stretching force to produce a noticeable effect. But we can reduce the stretching force needed by proper selection of the metal. Thus, a long wire (A in the illustration) made of thin lead or solder, stretched tightly by a weight (B), will lengthen noticeably in an hour and may even break in a few hours or overnight.

It is important that the wire not be severely weakened at any point along its length because it will tend to break there rather than to creep. Therefore, avoid sharp bends in the wire at points of attachment.

Cans of food are excellent weights and provide a wide range from which to select. Small bricks may also serve as

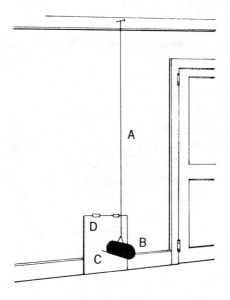

weights. But in all cases, keep a carefully supported box or other container a few inches below the weight so that no damage is done to the floor or to your foot if the wire should break.

A timer will prove handy. Set it to ring at regular intervals so that you can go about other business during the long intervals between readings.

TRY THESE INVESTIGATIONS

1. Obtain several feet of thin (about 1/16-inch diameter) solder wire and suspend a weight of about one pound from its lower end, as shown in the illustration. Attach a sharp point (pin, nail) to the side of the weight or to the wire, as shown at C and mark its position by pressing it into a sheet of paper (D) attached to the nearby vertical board or wall.

The weight causes the wire to "creep" and increase in length. Press the point into the paper every hour during the day to mark each new position.

After the wire breaks, remove the paper. From the markings measure the amount of stretch that occurred each hour. Does the creep occur at a regular rate—or do the distances of the stretch become greater and greater as time goes on?

A graph would be useful to show the changes.

2. Examine the nature of the break. If a micrometer is available, measure the thickness of the wire before it is stretched. Then measure it afterwards at the breaking point and also at other places along the wire. Does the thickness change at the point at which the wire breaks? Does it change elsewhere? It may be of interest to measure the thickness every inch along the wire.

3. Obtain solder wires of different thicknesses, but of the same composition. How is the rate of creep affected by the thickness?

4. What effect does increasing the weight have on creep? Test weights ranging from an ounce to 10 pounds.

5. What effect does greater length of the wire have on creep? Find out what happens with wires one, two, four, and eight feet long.

6. Solder is an alloy of lead and tin. Most solder wires are 50-50, that is, they contain equal percentages of lead and tin. Test 60-40 solder wire and other varieties. Is the rate of creep different for wires of different composition?

7. Test wires of other materials, such as lead, aluminum, and copper. Aluminum and copper are stronger than lead or solder so that you will need to use very thin wires and perhaps heavier weights. Or you may have to allow considerably more time—instead of a few hours or overnight—for your observations.

8. Construct a mechanism that will automatically record the amount of creep. One way to do this is to attach a cardboard disk to the minute or hour hand of an old clock, with a pencil attached to the weight in such a way that it presses against the slowly rotating disk. Then as the disk turns and the wire lengthens, the pencil mark on the disk will show exactly how the creep occurred.

An Investigation into Finding Latitude

BACKGROUND

If the earth were flat, we could describe the position of any place on it by referring to a numbered grid of parallel lines, like streets on a map. But the earth is round, so we cannot draw a grid of straight lines. We have to use numbered circles. The latitude circles run east and west. The longitude circles run north and south. In this investigation, you will find out how to measure latitude.

Imagine a circle drawn around the earth from the North Pole to the South Pole. This is a longitude circle, or meridian. Now imagine a series of circles (A, B, C, D, E, in the drawing below) running parallel to the equator, in the east-west direction. These are the latitude circles.

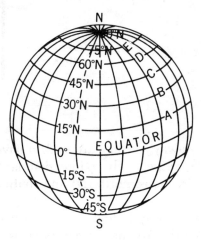

Each latitude circle is named for the point (measured in degrees) at which it cuts across the north-south longitude circle. This circle, like all circles, can be divided into 360 degrees. The starting point, 0°, is defined as the point at which the longitude circle intersects the equator. Thus, the equator is the circle of 0° latitude. The poles are a quarter

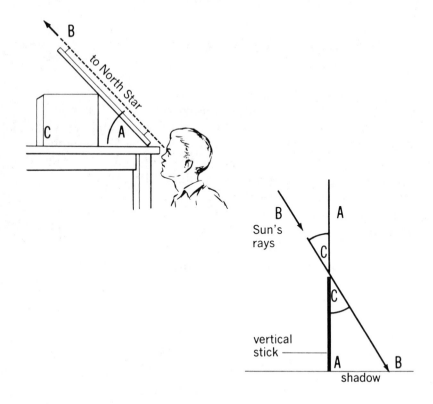

of a circle (90°) away, so they are said to have latitudes of 90°N and 90°S, respectively. (Note that latitude is measured in both directions, north (N) and south (S), from the equator.)

There is a simple way to find the approximate latitude of any place on earth. All that is necessary is to locate the North Star in the sky and measure its angle above the horizon (angle A in the diagram at the left, above), using a protractor. For example, at the equator the North Star always appears directly on the horizon; that is, its angle above the horizon is 0°, and the latitude of the equator is 0°.

Similarly, at the North Pole the North Star is overhead. The angle between the horizon (level) and the North Star (vertical) is thus 90°, and 90° is also the latitude of the North Pole. In the same way, at every other place on earth the latitude is equal to the value in degrees of the angle formed between a line drawn from the North Star and a line drawn to the horizon.

Another way to measure latitude is to measure the angle between a vertical stick or wall (A in the figure at the right) and the direction of the rays of the noonday sun (B) on the day when winter changes to spring (March 21), or the day when summer changes to fall (September 23). This angle (C) is equal to the latitude.

For example, at the North Pole on March 21, the sun is just beginning to rise above the horizon for six months of daylight. In this situation the angle between the vertical (overhead) and the sun (on the level horizon) is 90°. The latitude of the North Pole is 90°N.

At noon on March 21 and September 23, the sun is directly overhead at the equator. The angle between the vertical (overhead) and the sun is then 0°, which is the latitude of the equator. This method of finding latitude works at all places on earth.

TRY THESE INVESTIGATIONS

1. Determine your latitude by measuring the angle of the North Star above the horizon.

2. Determine your latitude by measuring the angle between the vertical and a line to the noonday sun on March 21 or September 23. (If you measure a few days before or after these dates, it will not make too much difference, unless you want to be very accurate.)

It is unwise to view the sun directly because your eyes may be damaged. The arrangement shown in the diagram on page 96 will be effective for making a measurement. Hammer a nail into a board, as shown in the diagram. Place the board on a level surface in a north-south direction. Mark the positions of the shadow of the nail on the board until it

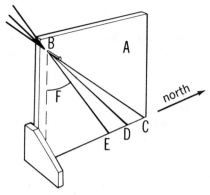

reaches its steepest position, when the sun is highest in the sky. Then, with a protractor, measure the angle between the vertical and the direction of the sun's rays (angle F).

3. The noonday sun is higher overhead in summer than it is in winter. Chart how the position of the noonday sun (that is, its angle with the vertical) changes with the seasons.

4. If you have kept your chart carefully, you will note that on June 21 the direction of the noonday sun is 23½° higher than on March 21. On December 21 it is 23½° lower than on September 23. Use this information to determine your latitude on the first day of winter or the first day of summer.

5. Write to someone who lives in a town north or south of yours. Arrange for both of you to make measurements of the angle of the North Star or of the direction of the noonday sun on March 21 or September 23. Determine the difference in latitude between your towns. Check with a map.

6. Improve the accuracy of your measurements. Keep these points in mind:

    a. We do not set our clocks by the local sun time, but according to the time zone. Thus, "noon," when the sun is highest overhead, may be from about 11:15 A.M. to 12:45 P.M. on your watch. And perhaps your state is on Daylight Savings Time for part of the year. Thus, you can't go by clock time in determining when the sun is highest in the sky (true noon). It will be necessary to watch a shadow of a stick to note when it is the shortest. This will occur at true noon, when the sun is highest in the sky.

b. When measuring the angle of the North Star above the horizon, keep in mind that this star is not located exactly above the North Pole, but revolves in a small circle around true north. If you want to be very accurate, make two measurements about twelve hours apart and average the two.

7. Use the method of finding true noon to find the north-south direction in your town. Watch the shadow of a vertical stick between 11:00 A.M. and 1:00 P.M. Mark the end of the shortest shadow. Draw a line from that point to the bottom of the stick. That line is north-south.

Does this direction agree with north as shown by magnetic compass? If not, find out why.

The investigations in this chapter all had the common characteristic that simple measurements were involved. In all of them, measurements of length or distance played a major role. But in each case, the procedures for making the measurement were different—and also meant different things. In the "honey clock," a change of distance as the container rolled down the board was used to indicate time. The measurement of a small distance on a photograph of the moon was interpreted in terms of very large diameters of craters. In the investigation of creeping metals, the measurement of length of a wire provided information about the strengths of metals and other properties. In the investigation of latitude, measurement of lengths of shadows or of angles gave basic information about one's location on earth.

You can see from this variety of simple measurements why they are so useful. Our minds are capable of manipulating these measurements so as to apply them to a wide variety of information-gathering purposes. In this way, we can find out much more about our environment than we could without the measurements.

CHAPTER 9

Combinations

of Measurements

In the previous chapters measurements in each investigation were of a rather simple nature. Rulers, clocks, and protractors were the major measuring instruments. In the main, devices were used in a straightforward manner.

Today a bewildering variety of measuring instruments is available to scientists. Scientists measure distances with micrometers, laser beams and radar, and other devices, as well as with rulers. The velocity of a star may be measured with a combination of a telescope, a spectroscope to produce a spectrum of the star's light on a photograph, and a microscope to measure small shifts in the spectrum lines.

One characteristic of such complex measurements is that they generally involve combinations of measuring instruments. Mathematical calculations are often required to obtain the desired measurement. Various techniques are necessary to prevent gross errors and to increase accuracy.

In this brief survey of the subject, it isn't feasible to suggest investigations in which complex instruments and calculations are utilized. But you can obtain some understanding of the basic processes of measurement by performing some relatively simple investigations with readily available materials.

An Investigation into the Strength of Magnets

BACKGROUND

You know that some magnets are strong and can lift fairly heavy weights. Other magnets are so weak that they can barely lift a small nail.

Magnets owe their strength to still smaller magnets inside them. Many tiny magnetic grains, called *domains*, pull together to cause the larger pull of the whole magnet.

As you know, every magnet has two poles, north and south. The domains also have such poles. But suppose that some of the poles of the tiny magnets inside the big magnet point helter-skelter rather than all in one direction. In that case, some tiny magnets pull while others push, and their effects may cancel each other out. Little or no magnetic effect then shows up outside the material. On the other hand, if all the little magnets line up and point the same way, they pull together and make the magnet strong. Therefore, one reason magnets differ in strength is the way their tiny magnetic grains (domains) are lined up.

Different magnetic materials also have different magnetic strength. The tiny grains in some magnets are strongly magnetic. In other materials, they are weakly magnetic. This factor accounts for the differences between some magnets.

TRY THESE INVESTIGATIONS

1. Measure the strengths of several magnets by observing how many small iron objects (paper clips, washers, nails, bolts, nuts, etc.) they can lift. Dip the end of the magnet into a pile of such objects, gather up as many as you can, deposit them elsewhere in a pile, and count the number of objects as the measure of the strength of the magnet. If one

magnet lifts twice as many clips as another, it is approximately twice as strong.

2. A more accurate method of measuring the strength of a magnet (as in the illustration on the left) is to use it to lift an object, such as a small empty (but clean) food can, through several layers of thin card or paper. Keep adding thicknesses of paper until the magnet no longer lifts the can. If another magnet lifts the can through more thicknesses of paper, then it is stronger.

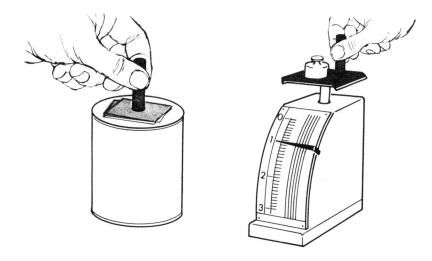

The accuracy of this test for weak magnets may be increased by using a thin, flat disk of iron as the object to be lifted and by making separators out of small, thin paper sheets that are all cut to the same size.

3. A small spring balance may also be used to measure the strength of magnets. Attach a flat disk to the bottom of the hook of the spring balance and pull it down with a magnet. The reading for the greatest stretch of the spring before it breaks away from the magnet is a measure of the strength of the magnet.

Postal scales may also be used. In most cases the platform of the postal scale is made of iron, and it is possible to pull up the platform directly with the magnet. Since the reading of the scale starts at zero, pulling it upward would cause a reading of less than zero, and there are no markings for such positions of the scale. Therefore, first load the scale with a weight and then observe how much less the reading is when the magnet pulls up on the platform. The arrangement is shown in the illustration on the right. Take the reading before the magnet pulls and again just before the platform pulls away from the magnet as you slowly pull upward. The difference between readings is a measure of the strength of the magnet.

This method works best if the weight used is just about enough to permit the magnet to pull the platform up to about the zero reading before it lets go. However, this condition is not absolutely essential.

An equal-arm balance may also be used in a similar way. If used, arrange it so that the addition of weight to one pan eventually makes it break away from the upward pull of the magnet on the same side.

4. How does magnetic strength change with increased distance of the magnet? The postal scale method described above is convenient to use for this investigation. It is essential that the platform of the scale be made of iron or that a flat iron object is attached to the surface. Cut about fifty squares (approximately 1½ inches x 1½ inches) of magazine cover paper (all of the same material). Place several squares in a neat pile on the center of the platform of the postal scale. Hold one pole of the magnet on the pile so that it attracts the platform of the scale through the paper. Because of the extra distance caused by the separation of the magnet from the platform, the attractive force is less. The greater the number of thicknesses of paper, the greater the distance separating the magnet from the platform, the less the attraction, and the less the force required to separate the magnet and the platform.

Make a record of the readings and make a graph to show how magnetic force decreases with distance.

5. Is magnetic force reduced with increased distance in

the same way for different magnets? Test a flat disk magnet and a long bar magnet. Make a graph of results for each. Is the shape of the graph the same? Is there some definite relationship between magnetic force and distance?

6. The method described above may be used to compare thicknesses of paper. Suppose that twenty-five sheets of one type of paper reduce the magnetic attraction as much as fifty sheets of another. Paper has no noticeable effect on magnetism. Therefore, we may conclude that the stack of twenty-five of one type of sheets is as thick as the stack of fifty of the other. Therefore, each sheet in the first stack is twice as thick as each sheet in the second.

Compare thicknesses of pieces of paper in this manner. If a micrometer is available, measure the thicknesses of the papers and verify your previous findings.

7. Does the way in which you hold a magnet affect its ability to lift iron objects. Use any of the previously used methods of measuring magnetic strength to find out.

Of particular interest in this respect is a disk-shaped magnet. Test its strength when the entire flat face of the magnet is in contact with an iron object being lifted, and again when the round edge is in contact.

An Investigation into Direction Finding by Radio

BACKGROUND

Do you have a portable radio of the type that has no outside antenna? If so, you have probably noticed that for each station the sound is loudest if you turn the radio to a certain position. If you rotate it a quarter turn (90°) past this position, the sound dies down quite a bit or disappears completely. You can make use of this observation to determine the direction of the radio station and even to locate its position.

Think of the way a water wave moves when a stone is thrown into a quiet lake. Two people standing in the water as the wave passes can easily locate the place where the

stone hit the water if each one faces in the direction from which the wave is coming. Imagine two lines drawn in these two directions, as shown. The place where the stone hit the water will be the point where the lines meet.

A radio station broadcasts a wave that is *spherical,* or ball-shaped. The round waves move outward in all directions from the station at the center. Some portable radios pick up the signal by means of a small coil-shaped antenna inside the set. In some of these sets the coil is so placed that the set sounds loudest when it faces the station. Other sets sound loudest when the long edge of the case points toward the station.

Here is a simple way to discover which situation applies to your set. First find out where a particular radio station is located. With an actual trial you can see whether it is the face (longer edge) or the side (short edge) of your portable set that points toward the station when the sound is loudest.

TRY THESE INVESTIGATIONS

1. Place your portable radio at the center of a level spot

on the ground. Tune in a station. Rotate the set until you get the loudest sound. Mark on the ground the direction of the radio station.

If you keep rotating the set, you will find that there are two opposite directions that produce the loudest sound. Your set cannot distinguish between these two directions. How, then, can you find out the true direction of the radio station?

Listen to the station for a while until the name of the city is announced. In most cases you will know if the city is north, south, west, or east of your location. If you don't know, a map will provide the information. You can then select the proper direction for the radio station.

Do this for a number of stations. Mark the directions on the ground. It might be interesting to make a permanent direction guidepost.

2. Locate north, east, south, and west by tuning in a station from a city whose direction you can find on the map. Mark these directions on the ground or in some other way. Check your measurements by means of a magnetic compass.

3. Airplane navigators use radio direction finding as one method of locating their positions on a map.

Imagine that you are in an airplane traveling above your town at a time when the ground is covered by fog. Find the directions of two radio stations and draw these directions on a map from the known location of the stations. The lines should meet at the location of the airplane (your town). Do they?

4. How can you improve accuracy of measurement? Here are some suggestions:

a. Locate the direction of weakest sound, rather than loudest.

b. It is a good idea to rotate the radio back and forth around the position of the weakest sound. You can then determine the direction of the station more accurately.

c. Attach a long stick to the side of your radio set. The extra length makes it possible to determine direction more accurately.

d. Use large maps rather than small ones.

An Investigation into Transparency

BACKGROUND

When you look through a sheet of glass, the view you get may seem as clear as without the glass. But actually a small portion of the light energy—about 5%—is absorbed in a clear, dust-free glass, while about 95% goes through. This 5% loss is usually not noticed, so it may seem to us that the glass has no effect on visibility.

But if we look through two sheets of glass, the loss is greater; with three, even greater; and with four, greater still. Eventually, the loss will reach 50%; in other words, half the light will be lost. At that point the reduction in brightness through the glass becomes quite noticeable.

The amount of loss of light energy coming through several layers of glass may be measured with a photographer's light meter. In such meters the markings are usually arranged so that each number on the scale represents twice the amount of light of the next lower reading.

In some light meters, the dial is adjustable so that any number can be set to correspond with the position of the pointer. Then suppose the light energy is reduced to half as much. In that case the pointer moves down the scale toward the "no light" position and stops at the next number.

The next lower number may not seem to represent half the amount of light, but it usually does. For example, the numbers may be in this sequence; 2, 2.8, 4, 5.6, 8, 11, 16, 22. If the pointer is at 2.8, this represents twice as much light energy as 2; 4 represents twice as much as 2.8; and so on. Many meters, however, operate in a different way. If you are in doubt about how to tell when the amount of light energy is half as great, read the instructions for the meter or ask a photographer.

Now suppose you interpose a number of sheets of clear material between the light source and the light meter. (Let's use plastic for our illustration because it is safer than glass.)

Each sheet absorbs a certain percentage of the light energy, and the reading of the meter is reduced. Eventually, as you interpose more sheets, the reading for amount of light energy will represent one half as much as before. If you count the number of sheets that caused the reduction to one half the amount of light energy, it is then possible to calculate the light-absorbing effect that can be assigned to each sheet.

Let us suppose that one sheet of plastic reduces the intensity of light by 10%. Most people would guess that it would take five sheets to reduce the intensity to one half (10% × 5 is 50%). But that is inaccurate. We can see why by working it out, adding one sheet after another. Suppose we have 100 units of light energy at the start. If the first sheet absorbs 10%, then it takes away 1/10 of the 100 units, or 10, leaving 90 units of light energy. Now there are only 90 units of light energy reaching the second sheet. Therefore, 10% of the light, or 1/10 of 90, or 9 units, are absorbed, leaving 90 minus 9, or 81 units, to reach the third sheet. The third plastic sheet absorbs 1/10 of 81, or 8.1 units, leaving 81 minus 8.1, or 72.9. In the same way, the fourth sheet takes away 1/10 of 72.9, or 7.29, or approximately 7.3, leaving 72.9 minus 7.3, or 65.6. The fifth sheet takes away 1/10 of 65.6, or 6.56, or about 6.6, leaving 65.6 minus 6.6, or 59.0. Actually, it will take seven sheets to reduce the light to less than half.

It is clear that mathematical skill is an important element in this particular investigation.

TRY THESE INVESTIGATIONS

1. Illuminate a white sheet of paper with a strong light. Point a light meter at it. It would be wise to keep the light meter stationary on a small box so that it is raised off the table and centered at the sheet without casting a shadow. Note the reading of the light meter. Move the meter, or the dial, until the pointer is exactly at one of the marked positions on the dial. Then interpose a pile of plastic sheets in

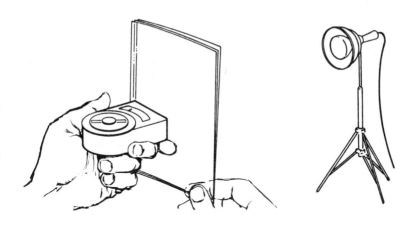

such a way that the entire light-sensitive area of the meter is covered. Keep adding sheets until the pointer reads half the amount of light—generally one marked division closer to the starting position of the pointer.

Count the number of sheets that reduced the light energy to one half. Calculate the percentage of light absorbed by each sheet.

2. Test various kinds of transparent or translucent sheets in this way—plastic bags, candy wrapper plastic, layers of clear gummed tape, etc. How do they compare in transparency?

3. Measure the light absorption of various liquids. Use a deep rectangular dish or bottle to hold the liquid and shine a light through the bottle toward a light meter. Compare the reading with that obtained without the liquid.

Investigate how greater distances of passage of light through liquids affect transmission of light.

4. Test the effect on transparency of one, two, three or more drops of colored dye in water.

5. Does dust or dirt on a transparent material affect its transparency? If so, by how much?

An Investigation into Light Acting at a Distance

B A C K G R O U N D

The sun—about 93,000,000 miles away—furnishes the earth with light. What would happen if the earth were suddenly moved twice as far away from the sun—to a distance of, say, 186,000,000 miles? Would the same amount of light still reach it from the sun? Or would it receive half as much light? What do you think? It would be interesting to set up an investigation that would provide an answer.

Of course, we cannot move the sun, but we can represent it by using a lighted bulb. To represent the earth, we need something that is affected by the amount of light that shines on it. This could be a light-sensitive paper or photographic film. Or it could be a photographic light meter, as described in the previous investigation.

T R Y   T H E S E   I N V E S T I G A T I O N S

1. Place the light source at various distances from the light meter. What does the light meter read at a distance of one meter? At a distance of half a meter? At a distance of two meters? Record your results in a table.

Do you find half as much light at twice the distance—or some other relationship?

2. There is a kind of photographic paper known as "studio proof paper" that slowly turns dark when exposed to light. No development with chemicals is required.

The amount of darkening of the paper depends on the amount of light energy that falls on the paper. The greater the amount of light energy that falls on the paper, the darker the sensitive paper becomes.

This kind of photographic paper is available in photographic supply stores. In a darkened room, cut pieces of this paper approximately two inches by half an inch for testing. Attach these pieces of light-sensitive paper to the center of a larger piece of gray-colored paper.

Place several test pieces at different distances from a light source (1, 2, 3, 4 feet, etc.) so that they face the source. Turn on the light. Note the time. Observe the darkening surfaces of the pieces of studio proof paper. Note the time when each test sample seems to be identical in grayness with the background square. Record the times in a table. Determine from this information how the amount of light energy changes with the distance. Take into account the fact that if it takes twice as long for the paper to darken by the same amount, then the illumination must have been half as much.

3. Improve your accuracy. Here are some suggestions:

a. Keep the room as dark as possible to keep other light sources (lamp, windows) from interfering with your measurements.

b. It is difficult to be exact about the time the studio proof paper becomes as gray as the background. Perhaps it will be helpful to note the time at which both papers (light-sensitive) and background begin to appear about equally gray, and the time when the proof paper is definitely darker. A time midway between the two might be taken as the reading. Perhaps you can devise a better method.

Some practice may be required before you make improved measurements with this type of observation.

c. Take into account variations in sunlight from day to day. The lamp, too, may change in brightness over a period of time.

An Investigation into the Water-Holding Capacity of Soils

BACKGROUND

Water in the soil is important for many reasons. Plants depend on it for the moisture they need to grow, and it is the source of much of the water used by man. Of course, there are many different kinds of soil, from the rich black loam of the prairies to the dry, sandy soil of the deserts; and these types vary widely in the amount of water they can hold.

Scientists have calculated the water-holding capacity of various general types of soil, and you can read about them in reference books. But no book will give you, accurately, the water-holding capacity of the soil in your yard, garden, farm—or even in your window box. A separate measurement must be made.

TRY THESE INVESTIGATIONS

1. What is the water-holding capacity of a sample of soil in a nearby farm or in your garden?

Here is a suggestion. Fill a given measured container with soil. Add measured quantities of water until the soil is wet and the water is level with the top of the container. How much water did you add?

2. What is the water-holding capacity of different samples from the same area? Are your measurements the same for all samples? Slightly different? Or very different? If they differ, can you devise a method of calculation that might

provide an indication of water-holding capacity of soils in the area?

3. Test different types of soil. Do they differ in water-holding capacity? Does topsoil differ from subsoil? Does soil on level ground differ from soil on a slope? Does soil near a spring hold as much water as similar soil elsewhere?

4. Improve the accuracy of your measurements. One basic suggestion involves equalizing water content of the soil at the beginning of the measurement. For example, the soil in one part of a garden may have been recently watered while another was not. The water-laden sample would thereafter absorb less water. Dry the soil samples by heating them moderately in an oven.

In this chapter you have seen a variety of problems in which combinations of measurements and calculations play a basic role in the investigations. Different measuring instruments were used. The measurements often required interpretation to make them meaningful. Mathematics was required for some to make them useful.

The last one in the chapter—water-holding capacity of soils—seems the easiest in concept—yet is actually more difficult than the others. The reason is that soils are not uniform—they may vary substantially from one scoopful to the next. Thus, we cannot simply state one number for all the samples that is true for all. But we can state an average.

Does this average remain the same if we duplicate the measurements for a new set of samples? We cannot be sure: scientists have developed special "sampling" techniques that increase the likelihood that representative measurements are made. Such sampling is of basic importance whenever wide variations occur in the phenomenon under investigation. Public opinion polls illustrate the problem. Back in 1935 a national magazine, *The Literary Digest*, took a poll of presidential preferences and concluded that Landon would win over Roosevelt. Instead Roosevelt won by a landslide. The magazine did not have a representative sampling of the population. They had taken the poll by telephone. At that time, only those who were relatively well off had telephones. Thus, the sampling was not typical of the population as a whole, and the extension of the results to the entire population gave incorrect results. This error was perhaps disastrous for the magazine. In any event, it went out of business soon thereafter.

Whenever there is wide variation in a subject under study, as was the case with the investigation of soil, then it becomes difficult to sort out the effect of different situations and factors involved. This is one reason why scientific studies in the fields of economics and sociology are often more difficult than for chemicals, forces, stars, or other subjects of scientific study that deal with things. People are very complex and differ widely. As a result, investigations involving activities by people tend to be more complicated, with many errors due to omission or ignorance of hidden factors. However, even here, scientific techniques are gradually being applied, and much new knowledge of vital concern is accumulating.

In the next chapter, we turn to another basic aspect of the "methods of intelligence," the experiment. Proper design of the experiment for an investigation frequently determines the outcome. Some clues about designing an experiment will be provided by the study of several investigations that focus on the problem.

CHAPTER 10

Designing
an Investigation

Investigations have to be carefully planned to obtain the information one seeks. Frequently this plan includes one or more experiments. Sometimes the design of a suitable experiment is quite simple—for example, in an investigation to find out which of two routes requires the least traveling time. In that case, the main procedure in the experiment consists of traveling the different routes on different days, and at different times of day, and gathering data for a number of trips until one can reasonably come to a conclusion as to which generally takes less time.

At the other extreme, the design of an investigation can be very involved and expensive. For example, scientists wanted to know whether a particular X-ray source that they had located in the sky was very small or quite broad. Since X-rays are absorbed by the atmosphere, it was necessary to send a rocket with X-ray detecting equipment into space out of earth's atmosphere. The rocket was then positioned so that the detecting instrument pointed at the X-ray source in the sky. The rocket was also projected along a path that would bring the moon between the distant X-ray source and the detecting instrument. If the X-ray signals detected by the instrument dropped off suddenly as the moon came between rocket and source, this would indicate a narrow X-ray source. If the detected signals tapered off slowly, this would indicate a broad source.

Can you imagine how much time and effort went into the design of such an investigation; how much it cost; and how many scientists, engineers, and assistants such an experiment required to make it successful?

Of course, there are many investigations not so costly or time-consuming that can provide reasonable practice for learning about the nature of experiments. Some of these are suggested in this chapter. In each case, observe how the design of the experiment is determined to a great extent by the nature of the problem. But also note that important details and techniques, subject to choice of the investigator, play an important role in the final design. Frequently the design must be changed to accommodate different conditions.

An Investigation into Maintaining Direction

BACKGROUND

You may have read stories about people lost in the woods who travel in circles even though they think they are walking in a straight line. It is said that each person tends to veer either to the left or to the right, just as everyone is either left-handed or right-handed.

How true is this?

It is clear that only an experiment can answer this question. The design of the experiment must include people walking toward a distant goal without reference to any landmarks or to the sun or stars. It would be difficult to set up a situation in which people set off in the woods toward a distant goal on a very cloudy day when the sun is not visible in the sky. The person would have to be tracked to be sure he doesn't really get lost. The observer would have to spend a great deal of time, and it might be fairly expensive to transport people to a suitable site, and perhaps feed them, too. Can we design a *simplified* procedure that would supply suitable information?

One good way is to blindfold the person, thereby temporarily preventing him from seeing the sun or landmarks. Then he must rely solely on his bodily motions to guide him along in a given direction. The blindfold should be light-tight to prevent sensing direction from the sun. Now, this isn't exactly the same as a situation in the woods, but there are very similar elements, and what we discover in this situation may well be applicable to all. Furthermore, we can use any open area that is not too large in extent—a school yard or a large gymnasium would do. No expense is involved because a trial can be completed in your neighborhood in a few minutes, and a large number of trials could be completed in a morning or afternoon.

TRY THESE INVESTIGATIONS

1. Blindfold a person, point him toward the goal, and let him walk toward the goal about a hundred feet away on level, uniform ground.

Does he go to the right or left of the goal? Does he zigzag, or travel in a curving path?

Repeat this procedure with a number of people.

2. Can a person learn to head in a given direction? Remove the blindfold after each trial and let the subject see how he walked.

Do people differ in ability to learn to walk in a given direction?

3. Do some people veer less than others as they head toward a goal blindfolded?

4. Do children differ from adults in their ability to head toward a goal blindfolded?

5. Do girls differ from boys in ability to walk straight when blindfolded?

6. In a large, flat area, is there any tendency to complete a circle? Do people differ in this respect?

7. Does a slope in the land affect ability to walk straight?

An Investigation into the Color Preferences of Insects

BACKGROUND

Do insects have preferences as to color? If there are color preferences, do different insects differ in this respect?

Of course, one might simply put some colored surfaces in a suitable place and observe if insects move toward any particular color. But time and resources are limited. Therefore, the design of this experiment should attempt to increase the likelihood that insects would approach your equipment and be offered a choice involving color.

TRY THESE INVESTIGATIONS

1. Flying insects are attracted to flowers, from which some take pollen and others take nectar.

It would be interesting to put sugar (instead of nectar) in colored containers and see whether insects visit containers of one color in preference to others.

You might paint several similar plastic vials yellow, green, blue, red, black, and other colors. Partly fill them

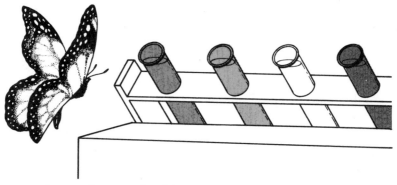

with a solution of table sugar. Perhaps you can support them in a rack like the one shown in the illustration. Set the rack outdoors and observe which containers the flying insects visit. Be sure to include one vial that is not colored, but white or clear. This vial serves as a control to see what happens if there is no color.

The sugar may also attract hummingbirds. If so, which color—if any—do they prefer?

2. Find the cone of an ant colony, with ants crawling to and from the opening in search of food. Near the colony, you might place bottle caps painted with different colors— or even colored squares on which you have placed powdered sugar. Watch to see which color the ants prefer.

3. Do butterflies differ from beetles and other insects in their color preferences?

4. Does the nature of the food affect the color preference? Test foods such as glucose, molasses, honey, bread, meat.

An Investigation into a Machine that Learns

BACKGROUND

When an animal like a cat is born, it already has certain abilities that are inherited. It doesn't have to learn how to breathe, open its mouth, or swallow milk. But many other activities have to be *learned*, often in combination with inborn, inherited acts called reflexes.

There are several different ways in which learning occurs. One of the simplest is by means of "trial and error." Suppose an animal performs an action. If the action has a pleasant or rewarding result, the animal is more likely to do it again. If the result is unpleasant or painful, the animal is less likely to repeat it. The animal soon learns which actions bring good results and which bring bad ones. If the punishment is sufficiently severe, even one experience is often enough. In this way it is able to adapt to its environment and escape harmful results.

One cannot escape the necessity of experimenting with real animals in order to find out how they learn. But often, it is extremely helpful to set up a *model* of the situation—without animals—and test hypotheses in a mechanical way, or perhaps by means of instruments. Models of this kind can often provide real insights into complex processes. In fact, we can even design "machines"—mechanical or electronic—that can duplicate some aspects of the learning process. Computers are excellent in that respect, but it is possible to set up model "learning" situations using nothing more than colored beads and boxes.

The learning situation should include some process that is to be "learned" by the "machine." There must be some way in which the "machine" can "express" its choices in each trial. The process should have an easily identified goal that indicates whether learning has occurred or not.

TRY THESE INVESTIGATIONS

Let us set up a familiar type of learning situation in which a real person, perhaps a friend, competes with a "machine" in learning how to win a simple game. For example, consider a game that begins with the placement of ten identical objects (coins, buttons, checkers, etc.) on the table. The players (human and machine) take turns removing objects. Each player is allowed to pick up one, two, or three objects

—but not more than three. The one who picks up the last object wins the game.

There is a simple method for winning, and we recommend that you try to find the secret yourself by playing the game with a friend. Then you can check your ideas with the explanation at the end of this section.

Set up the machine so that it can play this game, subject to assistance from you in carrying out its choice or moves.

We need some way for the machine to indicate its move. Here is a simple method. Put similar beads of three different colors—say red, green, and blue—in each of ten different jars. Mark the jars with numbers from one to ten. Put four beads of each color into each of the ten jars. The jars become the machine.

Of course, the jars can't do any selecting of beads, so they will need assistance from you. Suppose, after a given move by the machine's opponent, there are seven objects left on the table. You assist the machine to make its choice of the next move by selecting the jar numbered seven (which is the machine's move for this situation with seven objects on the table), closing your eyes, and *randomly* selecting a bead from the jar.

You must assign a meaning to the colors like the following: A red bead means, "Take one." A green bead means, "Take two." A blue bead means, "Take three." If you happen to take out a green bead from jar seven, this would mean that when seven objects are on the table, the machine "desires" to remove two. Of course, you will have to help by removing the two objects. In effect, you serve as the machine's agent in the purely mechanical process of selecting a bead and removing objects.

To illustrate, let us follow a typical game between your friend, whom we will call Harry, and the machine, with Harry making the first move.

Move 1. Harry removes two objects, leaving eight objects on the table.

Move 2. You randomly select from jar 8 a bead that turns out to be blue. This means, "Take three." You take away three objects for the machine, leaving five. Place the blue bead in front of jar 8.

Move 3. Harry removes three objects, leaving two.

Move 4. You select jar 2 (two objects left) and randomly pick a bead that is red. Red means, "Take one." You remove one object, leaving one. Place the red bead in front of jar 2.

Move 5. Harry removes the last object and so wins the game.

Now we come to the learning aspect. If the machine has lost, you want to "discourage" it from making such losing moves again. Remove all the beads that had been selected and placed in front of the jars. Put these beads in a dish. Note that in this process these jars now have a smaller number of beads that represent bad moves.

If the machine wins, you want to reinforce its winning moves. To do this, put back into each jar the bead that had been selected from it and, in addition, an extra bead of the same color. Each of these jars now has a greater number of beads representing winning moves.

You can see that as more and more games are played, the machine gradually loses beads that represent losing moves and gains beads that represent winning moves. Eventually, the chances are that only those beads that represent winning moves remain. If so, the machine has learned to play the game perfectly. It will then win whenever it goes first, and will also win most of the games—even when it goes second—against a human player who doesn't know the winning moves.

The machine has gone through a "learning" process very much like trial-and-error learning in animals. You can see that returning or adding beads to a jar is a kind of "reward" for having made a good move. Making a wrong move, which results in loss of the bead for that move, is a form of punishment. As the wrong-move beads are taken out of the jars, the chances of the right-move beads being selected are gradually increased, and the machine wins more and more often against a player who hasn't discovered how to beat the machine.

Try playing against the machine a few times. Make the best moves for yourself and play to win. How many games does it take for the machine to learn to win whenever it goes first?

Try playing against it without any plan. When it is your turn, just pick up objects at random as though you were a player who didn't know how to win. How many games now have to be played before the machine learns how to win? Does the machine learn to win faster or slower when you make random moves ?

What happens if all the right-move beads are left out of one container?

Try starting with different numbers of beads in the containers. Try "rewarding" the machine with two or three extra beads instead of one. How do such changes affect the machine's rate of learning?

✿   ✿   ✿

THE WINNING MOVES

We suggest that you do not read this section until you have given up trying to figure out the scheme for winning every game when you go first. It is more fun to find out for yourself. But, if you can't figure out, here is the plan for winning.

If, near the end of the game, you leave one, two, or three objects on the table, then your opponent wins the game by picking them up, leaving none. But if you leave exactly four, then your opponent must lose because whatever he takes (one, two, or three), you can pick up the rest and win. Therefore, four is one of the key numbers to leave after a move. In the same way, eight is a key number because if you leave eight objects, whatever your opponent takes (one, two, or three), you can take the right number (three, two, or one), to leave exactly four. Then he must lose.

The key numbers (the numbers of objects to leave after a move) are zero, four, and eight. We include the zero because that is what you leave on the last move and win the game. You can see that if there are more than ten objects to start with, the key numbers are zero, four, eight, twelve, sixteen, etc. Note that the main key number, four, is one more than three, the greatest number of objects you are allowed to pick up. If the rules were changed to allow at most two to be picked up, then the key number would be three. If at most four could be picked up, the key number would be five. Multiples of these key numbers would then represent the winning moves.

o   o   o

An Investigation into the Life Span of the Water Flea

BACKGROUND

Daphnia, the water flea, is a splendid laboratory animal. It can be grown readily, its young are produced in good quantity, and it is easily examined—even with a hand lens.

This insect is excellent food for tropical fish and can be obtained in pet stores that specialize in aquarium supplies. Most small ponds will also yield some *Daphnia*. To grow this organism, get a gallon or so of pond water. Crumble into the water a piece of hard-boiled egg yolk about the size of a marble. The egg yolk will supply food for bacteria, which, in turn, will become a supply of food for small one-celled animals called protozoans. *Daphnia* will feed on most microorganisms.

Now, suppose you have some *Daphnia*. How do you go about measuring its life span? It is difficult to tag or otherwise identify a given member of the species, and it is not reasonable for you to watch it all the time. Furthermore, when you get one, is it young or old? And how old? Clearly, special experimental techniques must be developed to find out how long a water flea lives.

TRY THESE INVESTIGATIONS

1. As soon as you have a supply of *Daphnia*, use a powerful magnifying glass to examine the animals for presence of eggs. They appear near the rear of the organism, about two to fifteen in number, as shown in the illustration.

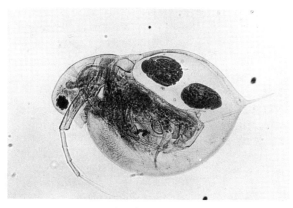

Carolina Biological Supply Company

No doubt you will find a good number of *Daphnia* with eggs. But, if you should make a mistake, there is no great harm done because these *Daphnia* will simply not produce young.

With a medicine dropper, isolate an adult with eggs in a vial half-full of the medium in which the *Daphnia* are growing.

Examine the vial regularly with a hand lens until you see that the young have been born. With a medicine dropper, isolate a young animal in a new vial and watch it grow. How long does it live?

One obvious complication in this plan is the fact that individual specimens differ. Consequently, it would be wise to isolate a number of individuals. Be sure they have adequate food and other conditions for life. It may then be possible to determine an average life span for the organisms.

2. Investigate portions of the life span. For example, how long does it take for a generation of newly produced eggs to grow to maturity and produce a new generation of eggs?

3. Do water fleas produce eggs faster at lower temperatures? At higher temperatures?

4. Does constant light or darkness affect the life span of the organism, or the time between egg-laying cycles?

5. Under what conditions is the life span of *Daphnia* the longest?

Daphnia can reproduce by parthenogenesis—a process in which mating is not necessary. You may wish to read more about this process and investigate it in *Daphnia*.

In this chapter you have observed several different designs for experiments. Each design was different because each situation was different. But there are certain elements in common.

First, note that simplification of the situation was a key element in all. In the investigation of direction, we performed the experiment in a school yard rather than in the woods. When testing color preferences in insects, we set up sources of food in colored containers, all in the same place, rather than selecting differently colored flowers that may be far apart and difficult to observe. In the investigation of learning, we used colored beads and containers instead of real animals. In the investigation of water fleas we kept them in small vials instead of studying them in a wide-open pond.

The purpose of this simplification was to make observations easier. Time and expense are vital elements in any investigation. An investigation that takes a few days, weeks, or even months is far more practical than one that takes years. An investigation that costs nothing, or a few dollars, is far more suitable than one that costs thousands or millions of dollars. An investigation that requires one or two observers rather than tens or hundreds is much more likely to be performed. Of course, there are situations in which years, millions of dollars, and hundreds of people are required to obtain meaningful results. Simple techniques are just not adequate. In such cases, the experiment might be done if the problem is considered sufficiently important.

Of course, there is always some loss in simplification. One can never be absolutely sure that the simplification significantly alters the results that one would obtain in the original situation. But the simplification is often valid and frequently provides important information.

Note that most of the experiments contained some sort of "control" in which one of the situations was designed to exclude the factor under study. For example, in the investigation of color preferences of insects, there was one uncolored (or white) vial. Suppose that this vial was left out and you observed that more butterflies went to a red vial. Your conclusion might be that butterflies are attracted to red. But suppose the butterflies were attracted much more to the uncolored (or white) vial than to *any* of the colored ones. Your conclusion would now be very different. In this case it would be clear that color has a repelling effect and that a red color repels butterflies the least.

Can you find the controls in some of the other experiments? If not, can you find some in which controls were lacking and should have been included? Frequently the control is not too obvious because it is built into the basic situation. For example, in the investigation of walking in a given direction, an uncontrolled situation would be one in which you had each person being tested start from a different place in the woods, head for a different goal, at different times of the day, on cloudy or sunny days. So many uncontrolled factors are present that one couldn't be sure what part each factor plays in what happens. The blindfold, the level ground, use of the same starting point and the same goal, all serve to exclude confusing factors and simplify the situation to a point where conclusions are more likely to be valid.

The main point is that situations in the real world are usually very complex, with many factors operating at the same time. The experiment is more likely to produce observation that can lead to a conclusion if the factors are varied *one at a time*, not two or more simultaneously.

As an example, suppose that a doctor is investigating the effect of a certain drug on a disease. He gives the drug to two people with that disease and finds that both improve. Can he conclude with any degree of certainty that the drug is useful in treating that disease? Not at all.

First, the two people may have been improving anyway, regardless of the use of the drug. Second, people differ, and even if these two patients improved, others may get worse or die if given the drug. Many factors enter into the observations. For example, it is difficult for doctors not to be prejudiced in situations where they believe that a drug is useful. Doctors (and scientists) frequently see results that they *want* to see, especially where differences are small. Even the seasons could affect the results. Suppose that the drug is given in February. By April, the patients may feel better because spring has arrived. Was improvement due to the drug or to the coming of spring? Perhaps the doctor transferred the patient to a relatively pleasant hospital environment from a miserable hospital situation and then gave his patient the drug. Was improvement due to the drug or to improved surroundings?

To eliminate the variations introduced by many factors, proper design for the experiment would include the following features to provide safeguards against error:

1. Many people would be included in the study, not just a few. The effects of variations among people would then average out.

2. Two equal groups of people with the disease would be selected for the experiment. One group would get the drug, and one would not. *All other conditions for both groups would be the same.* Any significant difference in the course of the disease between the two groups could then be reasonably attributed to the drug. This situation exemplifies the use of a "control."

3. Giving a pill often has psychological effects, and people feel better. Therefore, the control group, not getting the particular drug under study, should be given harmless dummy pills, called placebos.

4. It is best if the doctors who developed the drug do not also judge whether the patients are improving; otherwise, their own prejudices in favor of the drug can play a role in the outcome.

These are just some of the factors that must be included in a real experiment.

Another basic aspect of the experiment is that its basic purpose is to generate observations that provide essential information. You *do* something and see what happens. It isn't an experiment if one just thinks about it, figures out what should happen, and lets it go at that. One can never know if the conclusion is true or false unless it is put to actual test.

Finally, note that in each case in this chapter the design of the experiment arises from the statement of the problem. Once the problem is stated clearly, it often suggests the direction for the experiment and the design.

However, for more complex problems, originality in design becomes a more important factor in success. Occasionally there will be a brilliant new design for an experiment that solves problems where others have failed. But these are few and far between.

It might be worth another look at the investigations in this chapter to see if you could improve on the design or perhaps come up with some brilliant new idea of your own.

CHAPTER 11

Scientific Proof

Perhaps the most prevalent source of error in scientific experiments is the human failing of wanting the experiment to work. If one has a hypothesis in which one believes, the desire to prove it is often so strong that it causes the observer to select only those observations that buttress his case and to subconsciously reject all adverse information.

For example, in 1946, a doctor reported in a medical journal that a type of substance known as an "antihistamine" was effective in preventing colds. How did he know? He noticed that sniffling was reduced by antihistamines. Since colds produced sniffling, it seemed reasonable to suppose that antihistamines might also prevent the onset of the cold. So he gave antihistamine pills to his patients, with instructions to take them if they felt a cold coming on. Sure enough, most of his patients reported that the pills were effective in preventing colds.

The doctor announced his findings in a medical journal. Shortly several drug companies produced and reportedly sold about $100,000,000 worth of "cold pills" in a year.

However, independent research was undertaken by various medical teams to check out the truth of the doctor's claim. We will leave to the end of this chapter the details of design of their experiments. But for now, we can state that all the researchers found that the doctor's claim was completely unfounded. He and his patients had *wanted to prove the hypothesis* and had therefore selected the information that supported it. No attempt was made to set up situations in which adverse results might be observed. No attention was paid to a study of the large number of variables in the situation—the many conditions that could affect the results in different ways.

In this chapter are featured investigations that contain simple experiments to illustrate various scientific procedures that are used to eliminate error. See if you can find the procedures in each one that reduce the possibility of error.

An Investigation into the Relationship of Heredity and Environment

BACKGROUND

Each living organism comes into the world with a certain structure derived from its parents (heredity). It develops in an environment to which its structure must adapt. This structure helps it capture food and escape from enemies and other causes of death.

The relationship between heredity and environment of the many organisms on earth is so complex that we can't study all aspects at one time. We must first break down the problem into simpler elements that can be managed more easily. Since there are many variables (different types of conditions that affect the results), it is best to select a simplified type of situation that can be managed. For one thing, it would be wise to first narrow the range of investigation to one type of living thing—a potato plant, for example—rather than all living things, or even several.

Now, within each species there are differences between individuals. Potatoes from one plant will differ in some respects from potatoes originating from another parent plant. Therefore, it would be of interest to narrow the range even more. Can we obtain a number of potato plants, all of which have *exactly* the same heredity—identical twins, triplets, etc.?

This happens to be easy with potatoes. New potatoes may be grown from the "eyes" (buds) of a single potato. Each of these eyes, when planted properly, may grow into a complete plant, and *each has the same heredity as potato plants grown from other eyes of the original potato.*

You can obtain potato plants of identical heredity as follows. Cut a white potato into pieces of equal size, making sure there is an "eye" in each piece. Make the pieces as large as possible since this provides more food for the growing plant. Now plant the pieces each in a separate pot, with good garden soil. Plant each piece of potato one inch below the surface, with the eye upward. (Each eye is a bud.)

Keep the soil moist—not wet. In a week or so you should have several potato plants.

TRY THESE INVESTIGATIONS

1. What is the effect of different *amounts of light* on the growth of the plants?

2. What is the effect of different *colors* of light—red, green, blue, etc.—on the rate of growth of the plant?

3. What is the effect of the *size* of the originally planted piece of potato on its future growth?

4. What happens if you add *fertilizer* to the soil for some plants and not to others?

5. How do the plants accommodate to water with different amounts of *dissolved table salt*?

6. Are the plants harmed by *confinement* in a closed container?

7. At what *temperatures* do plants grow best?

8. Is growth of the plant affected by the *length of night*? For example, what happens to growth to maturity if light is on all the time?

9. Extend the above investigations to geranium plants.

Cut off a tip two inches long from a branch of the geranium. This is called a *cutting*. Take off all but three of the leaves at the tip of the cutting. Plant it one inch deep in sand in a paper cup. (Be sure to punch three holes in the bottom of the cup to let water run out.) In two weeks the geranium should be rooted.

Take several cuttings from one geranium to be sure that you have plants of the same heredity. Since you have controlled one major variable, heredity, you can now more readily investigate variables in the environment.

An Investigation into the Effect of Table Salt on Plant Life

BACKGROUND

There's plenty of water in the ocean. Could it be used to irrigate crops on the land—say oats, wheat, beans, corn, oranges? Probably you know that land plants are not adapted for growth in salt water. On the other hand, seaweed, algae, and other plants in the ocean are clearly adapted for life in salt water.

Ocean water is a solution of different salts: sodium chloride (NaCl), magnesium chloride ($MgCl_2$), potassium chloride (KCl), potassium bromide (KBr), and many others.

One effect that salt water has on land plants can be studied with solutions of table salt, which has the chemical name, sodium chloride (NaCl). Solutions of sodium chloride can be made in various concentrations. Marine algae can then be grown in different concentrations of the solution.

The question is: What concentration of sodium chloride stops the growth of a given plant? Can different plants grow in different strong concentrations of salt? Which plants can grow in weak concentrations or in plain water?

You will need to know how to prepare solutions of different concentrations. Solutions are specified in terms of percentages. For example, in a 1% solution, the weight of chemical is 1% of that of the solvent (usually water). To make a 1% solution of sodium chloride, weigh 1 gram of the substance and dissolve it in 100 grams of water. Or we could add 10 grams to 1,000 grams of water. In all of these solutions the ratio of salt to water is 1:100.

The medium in which you plant the seeds often contains minerals. This introduces a variable condition that you should eliminate. One way to do so is to use an inert medium, such as vermiculite or perlite, which do not supply dissolved minerals. These materials are often available from science supply companies, garden supply stores, or even from stores that sell insulation for homes. Or if sand is available, wash it thoroughly in tap water (provided this does not also have a high mineral content).

TRY THESE INVESTIGATIONS

1. Determine the effect of different concentrations of salt on the germination of seeds. Plant bean, oat, and beet seeds —at least three of each—in vermiculite (or washed and dried sand) in small flowerpots or paper cups. Be certain each seed is planted at the same depth (say half an inch) and in the same position.

Water the seeds with the salt solutions. Different pots should be watered with solutions of different concentrations, say 0.0%, 0.1%, 0.2%, 0.5%, 1%, 2%, 5%, etc.

2. If you live near the sea, obtain some algae growing in a pool of sea water. Place approximately equal quantities of algae and sea water in large, flat enamel or plastic pans. Keep them in a place where sunlight is available. Dilute several pans of sea water by adding one, two, four, or more times as much water as the volume at the start. Try one pan with fresh water. What happens to the way the algae grow with different concentrations of sea water? Which concentration is the best (the optimum)?

An Investigation into Population Growth

BACKGROUND

You hear a great deal nowadays about the "population explosion" and the possible consequences if the number of people continues to grow at the present rate.

Important information about this problem can be obtained from studies of animal populations. Many such studies are under way about populations of birds, fish, and insects in various universities and laboratories. You can duplicate some of the elements of such studies with a simple investigation of population of fish in jars of different sizes.

You may obtain all the necessary supplies—fish, water plants, fish food, sand, and a pump—at local aquarium supply stores.

TRY THESE INVESTIGATIONS

1. Obtain several one-gallon jars. Prepare each for supporting fish by adding two inches of sand to the bottom and pouring in pond water to within one inch of the top. If pond water is not available, use tap water from which the chlorine has been eliminated either by boiling and then stirring (to put back oxygen), or by letting it stand for about three days. Add water plants to help supply oxygen to the fish. A

plant known as *Elodea* is preferred, but other types of plants that are available in stores can also serve the purpose. Four to six sprigs of *Elodea* are about right, and be sure that you use the same number for all jars.

These oxygen-supplying plants make it unnecessary to change the water frequently—not more than once a month. If an aerating pump is used, the change of water may be even less frequent.

Stock each jar with two pairs of guppies. Feed the fish the same amount and kind of food each day, at the same time, according to instructions supplied with the food.

Keep data on population changes as they occur. Does the population of fish in each jar tend to stabilize? Are similar results obtained for each jar?

2. What happens to the population in two-gallon containers? Three-gallon containers? Five-gallon containers?

3. Is the population of fish affected by the amount of food?

4. Is the population of fish affected by the number of water plants at the start? What happens to the population if no plants are used? One sprig? Two? Three? Five? Ten?

An Investigation into the Breaking Down of Vitamin C

BACKGROUND

It would be convenient if we could be sure that the vitamins in foods were still there by the time they reached the dinner table, but it is known that under certain conditions vitamins break down into their chemical components. Let's try to find out what these conditions are. Does cooking destroy vitamins? Does heat? Does light? Is there a temperature range above which a vitamin breaks down most rapidly?

One of the easiest vitamins to investigate is vitamin C or ascorbic acid, readily available in drugstores. It occurs in many foods, especially vegetables and citrus fruits. There is a reasonably good chemical test for it; the blue substance dichlorobenzenoindophenol (commonly called indophenol blue) is turned colorless by vitamin C. This chemical is usually available in school laboratories, and it may be possible for you to obtain the very small quantity needed for your investigation by discussing your project with a teacher at your school who may supply you with it. If not, you may purchase the chemical from science supply companies, but buy as small an amount as you can because even a few pinches of it will go a long way.

A suitable solution of indophenol blue may be prepared for a fairly large number of tests by dropping small quantities of the powder (a pinch or so at a time) into a two-quart or gallon jar until a clear blue solution is obtained.

If you plan to carry out a very large number of tests, it would be preferable to make up a standard solution. A 0.1% solution contains 0.1 grams of indophenol in 100 grams (or 100 milliliters) of water. This solution may be prepared by weighing .1 gram of indophenol on an accurate scale and mixing it with 100 milliliters (ml.) of water, measured in a cylindrical graduate. However, you could certainly get started with a solution that contains approximate quantities.

It is necessary to standardize the test. Begin by making a solution of 0.01% pure ascorbic acid. This solution contains 0.01 grams of the substance in 100 grams (or 100 ml.) of water. Since it will be difficult to weigh such a small quantity of the substance, it would be easier to make ten times as much solution by adding and mixing 0.1 grams of ascorbic acid in 1,000 ml. of water.

Now pour 10 ml. of the indophenol solution into a clear container, and with a medicine dropper add the ascorbic acid solution, drop by drop until the blue color disappears. Count the number of drops required.

Next, do the same with the solution to be tested—perhaps some grapefruit juice, lemon juice, or orange juice. In other words, add the juice to 10 ml. of indophenol solution, drop by drop, until all of the blue color disappears. Compare the number of drops of juice needed for the indophenol to turn colorless with the number of drops of the standard solution of pure ascorbic acid. This information can provide a measurement of the concentration of vitamin C in the grapefruit juice. For example, if 20 drops of grapefruit juice caused the indophenol to become colorless, while only 10 drops were required of the standard solution of ascorbic acid, then we could conclude that there is less ascorbic acid per drop of grapefruit juice—in fact 10/20, or 1/2, as much.

On the other hand, suppose one drop of grapefruit juice caused the indophenol to become colorless. Obviously it is a more concentrated source of vitamin C than the standard solution, at least 10/1, or 10 times as concentrated. In that case, to further refine the measurement, add 9 ml. of water to 1 ml. of grapefruit juice (making 10 ml. in all). Then the juice is 1/10 as concentrated as before. Now if 3 drops of the diluted juice turn the indophenol colorless, you may say that 0.3 drops of the original solution caused it to turn colorless.

TRY THESE INVESTIGATIONS

1. Is vitamin C ever found in milk?
2. Is vitamin C destroyed by boiling? By heating to 150° F.?
3. How rapidly does vitamin C break down under refrigeration? Is it destroyed by freezing below zero?
4. How rapidly—if at all—is vitamin C destroyed by light? (Or is that a temperature effect?)

Let us examine the procedures in these investigations in greater detail. In each of these investigations the situation was simplified by limiting the number of variables. For example, in the Investigation into the Effect of Table Salt on Plant Life, it was suggested that you use vermiculite on washed sand (provided the water did not contain a substantial mineral content). The purpose of this procedure was to eliminate the variable effect on the results of different amounts of minerals in the planting medium.

In the Investigation into Population Growth, the containers were all prepared in the same way, with the same amount and type of sand, water plants, and fish population. In other words we "controlled the variables"—not all of them, to be sure, but some of the main ones. We did not control the variable heredity of the different fish nor that of the plants. Perhaps, if we are not careful and put one container where it gets more light than another, or more of a breeze, or put it near a radiator where it gets more heat, then we introduce a variable that may or may not affect the results. The point is that when we observe a difference due to some factor under study, we are likely to attribute it to the factor we are investigating, rather than to the unknown variable condition that may actually cause it.

For this reason it is generally important to use a control situation—you might think of it as a standard situation—and keep all conditions except one the same for all other trials. For example, when we go to two-gallon jars for the fish population investigation, we put in the same number of fish and plants as for one-gallon jars. But this is still tricky, because there are still some variables that are not really the same. Should we also put in the same volume of sand (in which case it would not be the original two inches high, specified for the smaller jar), or should we put in more sand than before until it is two inches high? If we use a round jar for the one-gallon container, would square-cornered tanks for the two-gallon containers introduce a new variable? If we test the one-gallon container in January, would it be justifiable to begin the two-gallon container experiment in July? Such questions are not at all easy to answer, and scientists must be very wary of the pitfalls introduced by *any* change in a situation.

In reality, no experiment can be *exactly* the same as another. If we do it in the same place, then the time must be different. If we do it at the same time, then the place must be different.

Fortunately, most situations are not seriously affected by minor changes in time or place. In many cases certain factors are so important that they determine results despite minor elements that are different. Therefore, it is possible to obtain valid results by controlling the main variables. But one never knowns for sure whether the factors we are not controlling are distorting our conclusions.

Note in the Investigation into the Breaking Down of Vitamin C how we made use of a measurement with standardized tests to obtain meaningful results. A standard solution of indophenol blue was prepared to provide a controlled situation in which a quantity of vitamin C could be determined.

Such a test introduces an "objective" element into the investigation that avoids "subjective" distortions by people. For example, one could imagine an experiment to determine whether vitamin C is present in foods that are fed to experimental rats, and their state of health could be observed. If the rat remains healthy with a given food, we might assume that vitamin C is present. If the rat becomes sick, should we assume that vitamin C is not present? And how do we tell if a rat is healthy or not? By looking at it? By weighing it and performing medical tests? How do we know that the basic food consumed by the rat, or even the water it drinks, doesn't contain vitamin C? Perhaps sunlight or other conditions of life affect the way the rat utilizes vitamin C. Experiments of this type with living creatures are conducted, but they contain many more variables than the indophenol blue test, and it is therefore much more difficult to obtain valid results. As you can see, if one wants to find out how much vitamin C is in a food, it is better to use the objective indophenol test rather than a more complex one with living things.

You will recall, at the beginning of this chapter, the story of the doctor who misled people about the "cold pill." He meant well, but he was a poor scientist because he did not control the variables and permitted subjective estimates of his patients to determine his results. One group of researchers who checked his results tried his hypothesis about antihistamines with a large number of people divided into two groups. One group received antihistamines. The other group was not given antihistamines, but received instead placebo (dummy) pills. All were given medical examinations during the period of investigation. The doctors giving the tests were not informed which of the people they were testing were getting antihistamines and which placebos. After all, the doctors might be influenced by preference for the finding that antihistamines could prevent colds.

All people in the test were asked to state whether or not they thought the pills prevented colds. The great majority of both groups thought so. But the objective tests showed no significant differences between the health of both groups. The doctor was completely wrong. He failed to use proper scientific methods and so arrived at an incorrect—and harmful—conclusion.

A carefully controlled, objective experiment is often expensive and time-consuming, requiring teams of researchers and large numbers of subjects. But there is no other way to conduct many investigations. In your own investigations you must keep in mind the many pitfalls that are possible and take steps to avoid them. Your results are then more likely to be valid. But remember that your conclusions and those of scientists can never be considered absolutely certain. We shall discuss this important aspect of scientific investigation in the next chapter.

CHAPTER 12

Can You Be Absolutely Sure?

In the previous chapter you observed some of the pitfalls that confront the scientist as he works on a problem, develops hypotheses, and designs and carries out experiments to gather observations and data. As the observations and data accumulate, the scientist may be increasingly convinced of the truth of his hypotheses (or perhaps of their falsity). At any rate, there usually comes a time when he must declare himself, take a stand, and publish his findings.

All of this time he has been gathering *evidence* about the hypothesis. Any specific piece of evidence is not, by itself, *proof.* But when the evidence piles up and becomes overwhelming, a point is reached when scientists may consider a hypothesis, or set of hypotheses, proved. It then becomes incorporated into the accepted body of knowledge. Frequently it requires decades of research before the question is considered settled. Sometimes it is not settled for centuries, and perhaps never.

But proof is tricky. In the world of the scientist, there is no such thing as *absolute* proof. The scientist's proof is not like that of the mathematician. An abstract mathematician does not deal with real things, but with symbols that he manipulates according to rules that he creates. If a mathematician begins with a few axioms (assumptions), definitions, and rules, he can generally build from it a large number of mathematical statements that follow logically from the axioms, definitions, and rules. In that sense, a mathematical proof of a statement can be absolutely "true"—or perhaps it would be better to say, *valid.*

In the scientist's real world, the mathematician's symbols can be very helpful, but they cannot ever *prove* a real fact. Only actual observations can provide real evidence, and a body of evidence, no matter how extensive, cannot prove an accepted fact true beyond a shadow of a doubt.

For example, scientists once thought they had proved that atoms were indivisible. With the discovery of radioactivity at the turn of the century, atoms were found to break apart by themselves. Further investigation revealed a detailed structure within the atom consisting of electrons, protons, neutrons, pions, mesons, and other components.

Again, in the 1930's, scientists were convinced that if the nucleus was bombarded with fast particles, it could only capture a simple particle or eject one. No one believed it possible that a heavy nucleus could be split into several major pieces. Scientists thought they had proof. But all the evidence was of a negative kind. Nobody had succeeded in splitting a nucleus. When Enrico Fermi, in the 1930's, began to try to create new types of atoms by bombarding the heavy atoms of uranium with particles called neutrons, he, and other scientists, all kept looking for new atoms that were only slightly heavier or lighter than uranium. They could not conceive that their information was false and that a nucleus could be split. It took years of mysterious and contradictory observations before Hahn and Strassman got the idea that atoms were actually being split apart.

As a result of such experiences, scientists today are quite cautious about "proof" in the sense of absolute proof. Any accepted fact applies only to certain conditions and situations. Alter the conditions or the situation, and a fact may still apply, but then again it may not. The only way to find out is by actual observation and experiment.

To understand the notion of scientific proof more deeply, it will be helpful for you to perform an investigation of a type a medical scientist might do in discovering the cause of a disease.

An Investigation into the Cause of a Disease

BACKGROUND

If you have ever kept tropical fish in an aquarium, you have probably observed the disease of fish known as Ich (pronounced "itch"). A fish affected by Ich first develops small white spots on its body. As it brushes against the sides of the tank in an attempt to get rid of the irritation, the spots spread. Soon the gills are affected, and unless the fish is treated to eliminate the condition, it dies. The condition must also be eliminated in the water in which the fish are living. The disease rapidly spreads to other fish, although it does not affect mammals.

TRY THIS INVESTIGATION

1. How would you go about discovering the nature of the white spots? Is it a bacterial infection of some kind?

To find out what causes the disease, catch a sick fish from an infected aquarium in a small net and place it on a piece of wet cotton. Scrape off some of a white spot with a cotton swab, dab it in a drop of water on a microscope slide, cover the drop with a cover slip, and examine it under the microscope. You will find organisms like those shown in the illustration; their scientific name is *Ichthyophthirius multifilius.*

Ichthyophthirius is fairly easy to grow away from the fish it affects. It will multiply rapidly in a medium composed of about a tablespoonful of malted milk added to a glass of water that has been standing overnight. In about a week, the water should be rich with the organisms.

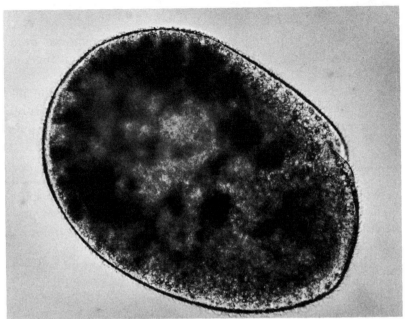

You can transmit the disease to a fish (such as a guppy) by filling a gallon jar three-fourths full of water from an aquarium, adding about a glassful of the medium with its organisms, and then placing the guppy in the jar. In about two weeks the white spots will begin to appear. You now have good evidence that the cause of the disease is a microscopic organism. But this is far from being proof.

2. How would you prove your hypothesis that the organism causes the disease Ich in fish? Scientists would demand that all the following postulates, laid down by the German scientist Robert Koch in 1882, be satisfied:

    a. The organism suspected of causing the disease must be present in every animal that has the disease.

    b. The suspected organism must be taken from an animal with the disease.

    c. The suspected organism must be carefully identified.

    d. The suspected organism must be grown in a pure culture, apart from the animal it infects.

e. The suspected organism, after being removed from the culture, must then be capable of transmitting the disease to a healthy animal.

f. The suspected organism, when removed from the animal it has just infected and grown again in culture, must be identified as the same organism as that in postulate c.

Design your own investigation to demonstrate how you would satisfy Koch's postulates for the disease of white spots in fish. Carry out this investigation and see if you can provide scientific proof of your hypothesis about Ich disease in fish.

In the course of this investigation, it is important to keep certain precautions in mind:

a. Take great care to design your controls carefully. For instance, it will not be possible for you to grow *Ichthyophthirius* in an entirely pure culture. The malted-milk medium will undoubtedly contain other organisms, such as protozoans and decay bacteria. Nevertheless, if you design your controls well, the evidence will point to one organism as the cause of the disease.

b. Be sure to cure your fish once the symptoms of the disease are established. While it is true that Ich is very common in fish, it is still your responsibility to keep the animals in your care free from pain. Fortunately, Ich can be easily cured if it is treated early. You can buy medicines for treatment of Ich at pet shops; one is Ich-Out, prepared by Longlife Fish Food Products, Denville, New Jersey.

c. Take care not to infect healthy fish in an aquarium from which you take fish for investigation. A net used in handling infected fish should not be dipped into the aquarium. Also, be sure a fish is cured before returning it to the aquarium.

One thing is very clear from this investigation: scientific

proof is difficult work. All of the skills of the scientist and lots of time must be applied. In this case your task is made easier because you have been told that Ich is the cause of the disease. All you are doing is verifying what others already know. But put yourself in the place of a scientist investigating an unknown disease. There are many possible causes: infection by organisms, chemical deficiency in food, defective chemical actions in the body, radiation, pollution, food poisoning, and many still only guessed at—as with cancer. As you can see, Koch's postulates apply to only one type of cause, infection by an organism. Similar rigorous conditions of proof must be met in all fields of scientific work.

CHAPTER 13

Generating Problems

Most people have problems without asking for them and are happy to be relieved of them. Scientific problems are different. The scientist looks for problems to solve. In fact, he devotes his life to finding problems to solve and then solving them.

Scientific problems are generated in various ways. To illustrate the way problems arise, read the investigations that follow and try to do some of them. In each case note especially the way the problem arises. In all cases, of course, the problem is stated or implied in some manner in this book.

It is of special interest in this chapter to note how problems arise in a natural manner as one digs into an investigation in any depth. Look for this feature of investigations as you read or do those described in this chapter.

An Investigation into the Action of an Antibiotic

BACKGROUND

Antibiotics, the modern "miracle drugs," are used by doctors to treat a wide range of diseases. Their ability to fight infections is dramatic, but like most powerful drugs, they also have other effects. For example, large doses of the antibiotics Aureomycin and Terramycin have been found to

148

speed up the growth of some animals—particularly mammals.

If these antibiotics affect the growth of multicellular animals, do you suppose they also affect the growth of microscopic one-celled animals known as protozoans? A typical microscopic animal that would be suitable for investigation is *Paramecium*, available from biological supply companies, or you can grow some yourself in jars.

Do antibiotics increase the size of paramecia? Do they, perhaps, increase the rate of fission? Perhaps they have no effect at all. Perhaps they have an effect we cannot predict.

You will need a supply of *Paramecium*. First, collect a gallon of water from a pond. Use it to fill eight pint jars. Crumble into each jar a bit of hard-boiled egg yolk, about the size of a pea. Keep the jars loosely covered in moderate light at room temperature. Within one or two weeks the water should be teeming with paramecia so that you can proceed with the investigation.

You will also need a supply of antibiotic. Since these are available only with prescriptions, you may wish to ask your teacher for a supply. It would also be well to obtain information. Suppliers of the antibiotics generally provide informational booklets that would be helpful in working with their antibiotics. Aureomycin is produced by Lederle Laboratories, Pearl River, New York. Terramycin is made by Charles Pfizer and Company, Inc., 235 East 42nd Street, New York, N.Y.

You will want to start with a known concentration of the antibiotic. An accurate weighing scale and cylindrical graduate will be required. You may find such instruments available in your school, or they may be purchased from science supply companies at moderate cost. Begin with a 0.1% solution made by dissolving 1 part of antibiotic in 1,000 parts of water; for example, 0.1 gm in 100 grams (or 100 ml.) of distilled water.

If you want to try the effect of a weaker solution, say

0.01%, you can make it from your stock solution. Here is one way of doing it. A 0.01% solution must contain 0.01 grams of the antibiotic in 100 ml. of water. If you want to calculate it, you will find that 10 ml. of a 0.1% solution of a substance contains .01 of a gram of the substance. To prepare a 0.01% solution, then, you need merely add 10 ml. of the 0.1% solution to 90 ml. of distilled water.

TRY THESE INVESTIGATIONS

1. Grow a culture of *Paramecium* in pond water. Add a sufficient quantity of 0.1% Aureomycin to double the volume of liquid in the jar. Then make a daily comparison of the number and size of paramecia in that jar with corresponding data for a control culture to which only distilled water (or pure rainwater) has been added.

How would you count the number? Here is one way. Thoroughly stir the medium in which the animals are growing, and then take out a 50 ml. sample. Under the microscope, count the number of paramecia in one drop of the sample. Do this for at least ten other drops. (More would be better. Why?) Calculate the average number of paramecia per drop.

Size may be determined with a reticle inserted into the eyepiece, which projects an accurate ruled scale against the image of the organism. Measure the size of at least ten paramecia and obtain the average.

2. Try the effect of different quantities of Aureomycin on growth of paramecia.

3. Test the effect of Terramycin and other antibiotics on growth of paramecia.

4. Test the effect of antibiotics on the growth of other organisms—for example, yeast.

An Investigation into the Speed of Germination of Seeds

BACKGROUND

In some farming areas growing seasons are short. It is therefore of deep interest to the farmers in those areas to obtain seeds that germinate rapidly and grow to maturity quickly. Or again, in areas with longer growing seasons, a fast-germinating and fast-growing plant might be planted twice in one season.

Well, then, put yourself in the place of an agricultural scientist confronted with this problem. What would you do?

Here are some techniques that will prove helpful in this investigation:

    a. Prepare a germinator. Place a piece of blotter in a clear plastic container. Keep it moist with a wick that brings in water from a jar. Invert a second container on top of the first. Be sure to keep the small jar full of water.

    b. Before you place the seeds in the germinator, soak them in water overnight. This will furnish the water

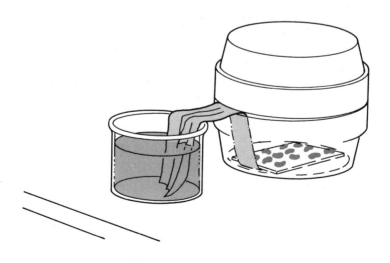

that the cells of the seed need to absorb before they become active.

TRY THESE INVESTIGATIONS

1. Test the effect of light or darkness on the speed of germination.
2. Test the effect of colored light on the rate of germination.
3. Test the effect of various high and low temperatures on the speed of germination.
4. Test the effect of various dilute solutions on the rate of germination.
5. Test the rates of germination of various types of bean seeds; various types of other seeds.

An Investigation into Vegetative Propagation

BACKGROUND

Probably you know that a willow twig takes root when placed in moist sand or in water. Part of a plant is thus capable of reproducing the whole plant in some situations. We call this process vegetative propagation.

You probably also know that an onion bulb can reproduce the roots and the shoot. A piece of a white potato that contains an eye can reproduce the entire plant. A root like a sweet potato can also produce a whole new plant. A cutting of a geranium, like the one in the drawing, can reproduce an entire geranium plant.

TRY THESE INVESTIGATIONS

1. Cut off, with a knife, a piece of geranium stalk about six inches long. Place the cut end in a jar three-quarters full of moist sand so that the cut end can be seen through the glass. In that way you can determine how fast roots form. How long does it take for roots to form?

2. Would a four-inch stalk produce roots? A two-inch stalk? A one-inch stalk? Is there a size (called a critical size) below which the stalk will not regenerate an entire plant?

3. Do stalks of different sizes regenerate at the same speed?

4. Are your findings also true of willow twigs? Will a thick twig regenerate faster than a thin twig?

An Investigation into the Effect of Aspirin on Keeping Cut Flowers Fresh

BACKGROUND

Aspirin is a drug prescribed by physicians for the relief of certain pains, particularly headache. Like most drugs, aspirin (or sodium acetylsalicylic acid) is generally beneficial in prescribed amounts and poisonous in larger amounts. However, it is commonly found in medicine cabinets. (*Caution:* Like all medicines, it should always be kept out of reach of young children.)

Recently, aspirin has been said to prolong the life of cut

flowers. The practice seems to be to dissolve one tablet in the water in which cut flowers are placed.

Does aspirin actually prolong the life of cut flowers?

TRY THIS INVESTIGATION

1. Cut some fresh flowers. Be sure to include several varieties, both wild and cultivated. The water in which your cut flowers stand, the temperature, the light conditions, even the kind of container, will need to be taken into consideration in designing the experiment.

2. What concentration of aspirin in solution is best for keeping cut flowers fresh?

An Investigation into the Effect of Light on Tree Leaves

BACKGROUND

As you drive along a tree-lined street, you may observe that street lights shine on some trees but not on others. Would such constant light on the leaves of some trees have any observable effect?

Should there be any effects? For one thing, a tree that is subjected to a longer period of light may undergo longer periods of photosynthesis. It may need more water. Its leaves may develop differently. We know, for instance, that trees grown in the shade generally have thinner, larger leaves than those of the same species grown in full light. Perhaps the leaves of trees exposed to constant light fall at a different time as winter approaches.

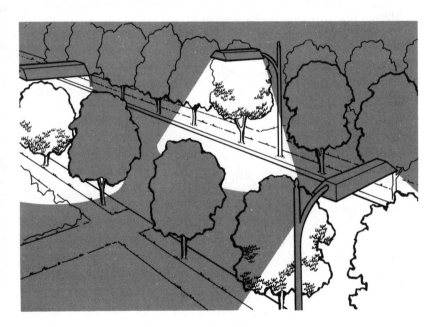

TRY THESE INVESTIGATIONS

1. Do similar trees under different conditions of light flower at the same time? Compare the flowering of trees that are distant from streetlights with trees of the same type that are illuminated by nearby streetlights.

2. Do trees under different conditions of light bear fruit at the same time?

3. Does the time when leaves fall differ for trees under different conditions of light?

4. Do trees under different conditions of light have different autumn coloration?

Perhaps you will be able to think of other questions.

As you think back over the investigations in this chapter, you will observe different ways in which problems for investigation arise. In the case of the investigation of an antibiotic, the problem begins with the observation that antibiotics speed up growth of mammals. It is quite natural to extend and generalize such a thought to, "Antibiotics speed up the growth of all animal life."

Of course, such a hypothesis is so broad that it demands that we break it down into smaller manageable hypotheses, such as: "Antibiotics speed up growth of fish" (or birds, or insects, or one-celled animals, or specific types of animals such as chickens, flies, or paramecia).

Each hypothesis also implies a problem. It says, "Prove me true or false." If we prefer, we can always state the hypothesis as a problem. "Do antibiotics speed up growth of fish?" (Or of any other organisms you can think of.) If this problem is solved and we find that one antibiotic does indeed speed up growth in a type of fish, then we can check a large number of additional problems with regard to other types of fish, other types of animals and plants, degree of concentration of antibiotic, effect of various conditions, etc. Each investigation adds to our knowledge; eventually we come to know a great deal about antibiotics.

Any new discovery has this quality of generating large numbers of hypotheses and accompanying problems. For example, when the first antibiotic, penicillin, was discovered by Fleming as a result of observation of a clear, germ-free area around a culture of the common mold known as *Penicillium*, a vast search for other germ-killing molds was begun. Aureomycin and Terramycin are two of many antibiotics that were discovered as a result of this search.

The search for antibiotics was pursued in different ways, depending on how the problem was formulated. In some cases, scientists formulated problems around specific molds. "Does mold A retard growth of disease-producing organism B?" In other investigations a vast variety of different kinds of soils were investigated to see if molds or other forms of minute life they contained have germ-killing powers. The motivation for much of this searching was practical—men wanted to cure disease. But some of the motivation was pure curiosity. For example, Fleming made his original observation and followed it through mainly because he wondered why a clear area appeared in his mixture, and not primarily because he was sure it would lead to useful antibiotics—although that thought might well have lurked in the back of his mind.

The Investigation into the Speed of Germination suggests another example of a practical source for problems to solve. The farmer has many real problems to contend with in growing his crops. He needs methods of overcoming crop disease, drought, insects, early frost, late thaws, and lots more. Since the world depends on food and therefore on the farmer, governments spend large sums of money on scientific research with the practical aim of solving agricultural problems.

But there are also industrial problems, consumer product problems, pollution problems, transportation problems, psychological problems, urban problems, education problems, military problems, and lots more—all stemming from needs felt by governments, businesses, or groups of people. These problems demand solutions, and problem-solving scientists of all kinds are therefore hired to work at finding solutions.

The waves of information emanating from the search for solutions to these practical problems of scientists often have important consequences in so-called "pure" science, where scientists work on problems that seem to be remote from practicality. For example, an electronics scientist at Bell Telephone, Karl Jansky, while investigating the source of some faint, static-like "radio noise," discovered radio waves coming from outer space. This opened up the highly important field of radio astronomy and has already contributed much to our knowledge about outer space. Practicality is not the direct motivation for these studies—although, no doubt, some day there will be results of direct use to mankind.

Another source of problems is indicated by the Investigation into the Effect of Light on Leaves. An observation of an everyday situation or event can generate problems for solution. If someone has basic knowledge in the field of botany and is an investigator, the observation of streetlights shining on some leaves and not on others would perhaps suggest problems for investigation. Here is a neat difference in conditions for some trees and not others—some get artificial light at night and others do not. If we were to design an experiment of this type for trees, it would be prohibitively expensive to set up. And here it is ready-made, in many of our city streets.

Long ago, such keen observations of everyday events were more likely to be productive of new knowledge. Today, it still does happen, but rarely, and generally in less important ways. Most events close to our lives have been thoroughly explored, and the big action is in the outer reaches where the same skill at generating problems from observations is more likely to be fruitful. A scientist studying cosmic ray tracks in a photograph may observe a track that looks different. If he pursues this observation, he may end up discovering a new type of particle. But he must first know his field well so that he can recognize the different track from the common ones.

And that, of course, is the main point. One needs to know a great deal about a subject before he can hope to contribute to it by originating new problems and solving them.

Are you to become an investigator?

It is a rich and rewarding life—whatever field of science you choose.

Here are you—very young and able. Here you are—ready to turn your attention to the kind of work you want to do. Here you are—ready to contribute to civilization.

This is altogether fine. You are needed.